Springer Tracts in Advanced Robotics

Volume 155

The Springer Tracts in Advanced Robotics (STAR) publish new developments and advances in the fields of robotics research, rapidly and informally but with a high quality. The intent is to cover all the technical contents, applications, and multidisciplinary aspects of robotics, embedded in the fields of Mechanical Engineering, Computer Science, Electrical Engineering, Mechatronics, Control, and Life Sciences, as well as the methodologies behind them. Within the scope of the series are monographs, lecture notes, selected contributions from specialized conferences and workshops, as well as selected PhD theses.

Special offer: For all clients with a print standing order we offer free access to the electronic volumes of the Series published in the current year.

Indexed by SCOPUS, DBLP, EI Compendex, zbMATH, SCImago.

All books published in the series are submitted for consideration in Web of Science.

Kyu Min Park · Frank C. Park

Collision Detection for Robot Manipulators: Methods and Algorithms

Springer

Kyu Min Park
Korea Institute of Science and Technology
Seoul, Korea (Republic of)

Frank C. Park
Seoul National University
Seoul, Korea (Republic of)

ISSN 1610-7438 ISSN 1610-742X (electronic)
Springer Tracts in Advanced Robotics
ISBN 978-3-031-30194-0 ISBN 978-3-031-30195-7 (eBook)
https://doi.org/10.1007/978-3-031-30195-7

This Springer imprint is published by the registered company Springer Nature Switzerland AG
The registered company address is: Gewerbestrasse 11, 6330 Cham, Switzerland

Series Editor's Foreword

At the dawn of the century's third decade, robotics is reaching an elevated level of maturity and continues to benefit from the advances and innovations in its enabling technologies. These all are contributing to an unprecedented effort to bringing robots to human environment in hospitals and homes, factories, and schools; in the field for robots fighting fires, making goods and products, picking fruits, and watering the farmland, saving time and lives. Robots today hold the promise for making a considerable impact in a wide range of real-world applications from industrial manufacturing to healthcare, transportation, and exploration of the deep space and sea. Tomorrow, robots will become pervasive and touch upon many aspects of modern life.

The *Springer Tracts in Advanced Robotics (STAR)* is devoted to bringing to the research community the latest advances in the robotics field on the basis of their significance and quality. Through a wide and timely dissemination of critical research developments in robotics, our objective with this series is to promote more exchanges and collaborations among the researchers in the community and contribute to further advancements in this rapidly growing field.

The monograph by Kyu Min Park and Frank C. Park is based on the first author's doctoral thesis. The contents are organized in six chapters and three appendices. The focus is on the key problem of physical human–robot interaction for collaborative manipulators with special emphasis on safety. Collision detection is the main thread, where model-based and model-free methods are keenly combined for enhanced learning of collisions and their effective handling.

A nice blend of original theoretical results with application to robot manipulators, this volume is a fine addition to the STAR series.

Naples, Italy
February 2023

Bruno Siciliano
STAR Editor

Preface

While there is no standard definition of what constitutes a collaborative robot manipulator, from the perspective of physical human–robot interaction (pHRI), one can imagine a robot capable of operating in dynamic and unstructured environments shared with humans (and other robots), physically interacting with their human operators on given tasks. It goes without saying that the safety of the human operator overrides any other performance considerations and that the most basic requirement to ensuring safety is the fast and reliable detection of collisions.

This book attempts to provide a concise, reasonably comprehensive (but not exhaustive) survey of collision detection methods that have been developed for robot manipulators. Beginning with a review of robot models and preliminaries on basic statistical learning methods, the book discusses the fundamentals of the collision detection problem including collision types and collision detection performance criteria, model-free versus model-based methods, and the more recent data-driven learning-based approaches to collision detection.

Special effort has been given to describing and evaluating existing methods with a unified set of notation, systematically categorizing these methods according to a basic set of criteria, and summarizing the advantages and disadvantages of each method. As far as we know, this book is the first to comprehensively organize the growing body of learning-based collision detection methods, ranging from basic supervised learning-based methods to more advanced approaches based on unsupervised learning and transfer learning techniques. For some of the learning-based methods that we have developed, step-by-step implementation details and pseudocode descriptions of the algorithms are provided. For collision detection performance criteria, we cover not only conventional criteria such as detection delay and the number of false alarms but also learning-related criteria such as generalization capability.

We hope the contents of this book will be useful to robotics engineers interested in endowing their robots with collision detection capabilities, particularly the algorithmic and software aspects of learning-based approaches. Whether it be for both research or commercial applications, or in settings ranging from industrial factories to laboratories conducting experiments on complex physical human–robot interaction tasks, we believe the book can help the reader choose and successfully implement the most appropriate detection method that suits their robot system and application.

Seoul, Korea (Republic of)
February 2023

Kyu Min Park
Frank C. Park

Acknowledgements The work in this book has been supported in part by IITP 2021-0-0268 AI Innovation Hub/SNU-AIIS, SNU Soft Robotics Research Center, SNU-IAMD, and SNU Engineering Research Center.

Contents

Acronyms

1-D CNN	One-dimensional convolutional neural network
CNN	Convolutional neural network
DOF	Degrees of freedom
F/T	Force–torque
FNN	Fully-connected neural network
FSR	Force-sensitive resistor
KKT	Karush–Kuhn–Tucker
LSTM	Long short-term memory
MCD-AE	Model-compensating detection using an autoencoder
MCD-CNN	Model-compensating detection using a 1-D CNN
MCD-OCSVM	Model-compensating detection using an OC-SVM
MCD-SVMR	Model-compensating detection using SVMR
MNN	Modularized neural network
OC-SVM	One-class support vector machine
pHRI	Physical human–robot interaction
PRAUC	Area under the precision–recall curve
RBF	Radial basis function
ReLU	Rectified linear unit
SRQ	Squared rational quadratic
SVM	Support vector machine
SVMR	Support vector machine regression
SWCVAE	Sliding-window convolutional variational autoencoder
ZMP	Zero moment point

List of Figures

List of Tables

Chapter 1
Introduction

1.1 Collaborative Robot Manipulators and Safety

A robot manipulator typically refers to a mechanism constructed by serially connecting rigid bodies, called *links*, by mechanical connecting elements called *joints*, forming an arm-like structure. Each of the joints is actuated by a motor that causes relative motion between adjacent links; by controlling the motors, the manipulator can be programmed to move and exert forces in desired ways. For a serial arm connected by either rotary (revolute) or linear (prismatic) joints, its total *degrees of freedom* (*DOF*) is simply the total number of joints. The majority of industrial robot manipulators today have a DOF of six or seven [1].

With their fast speed, high accuracy, and increasingly low production cost, robot manipulators are widely used in industrial settings, performing a range of repetitive or dangerous tasks in place of humans. Most robot manipulators today are only capable of executing commanded (pre-programmed or pre-learned) tasks, limiting their ability to perform more complex tasks and quickly adapt to changes in their environment. Humans on the other hand can think and make quick decisions, and precisely manipulate objects in unstructured and highly dynamic settings. For these and other reasons, collaborative manipulators hold the promise of simultaneously leveraging the capabilities of both humans and robots.

Collaborative robot manipulators must not only operate in dynamic and unstructured environments shared with humans, they must also physically interact with human operators. In the context of *physical human-robot interaction* (*pHRI*), the first and most important requirement of a collaborative robot is maintaining the safety of the human operator [2–4]. Toward this end, safety standards such as the Technical Specification ISO/TS 15066 [5], which specifies detailed safety requirements for industrial collaborative robots, have been established.

K. M. Park and F. C. Park, *Collision Detection for Robot Manipulators: Methods and Algorithms*, Springer Tracts in Advanced Robotics 155,
https://doi.org/10.1007/978-3-031-30195-7_1

1.2 Human-Robot Collision Handling

The primary considerations in safety are to predict and avoid potential risks, i.e., accidental collisions between the human and the robot, and if collisions do occur, to minimize injury to the human as much as possible. It is imperative that safety be addressed at multiple levels. At the design level, the mechanical structure of the robot should be designed to minimize injury when a collision occurs. To reduce the impact force, mechanical systems such as soft artificial skins that cover the links [6–8], variable stiffness actuators that switch to a compliant mode as soon as a collision is detected [9–11], and lightweight robots [12–14] have been developed.

At the control and planning levels, the robot should attempt to avoid collisions in the first place, e.g., planning collision-free trajectories [15–19] and real-time instantaneous avoidance of potential collisions [20–23]. When collisions do occur, strategies should be developed to handle, on a case-by-case basis, subsequent actions of the robot based on the nature of the collision, including determining whether or not the collision is intended or unintended, whether the collision is hard or soft, and other considerations that we address below.

The three levels at which safety can be addressed—mechanical design, collision avoidance, and collision handling—have different levels of expected injury level and range of applicability (see Fig. 1.1). Naturally, the risk of injury is very low if physical contact between the human and the robot can be avoided altogether. If a robot is designed to minimize the impact force resulting from a collision, injury levels can be considerably reduced (although not to the extent as avoiding collisions). Since collaborative robots are purposefully designed to cooperate with humans, physical contact is an integral element of pHRI, and it follows that a collision handling

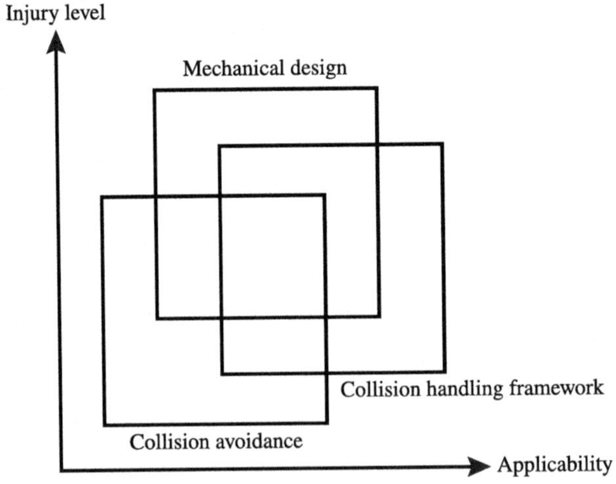

Fig. 1.1 The expected injury level and potential applicability of safety handling strategies

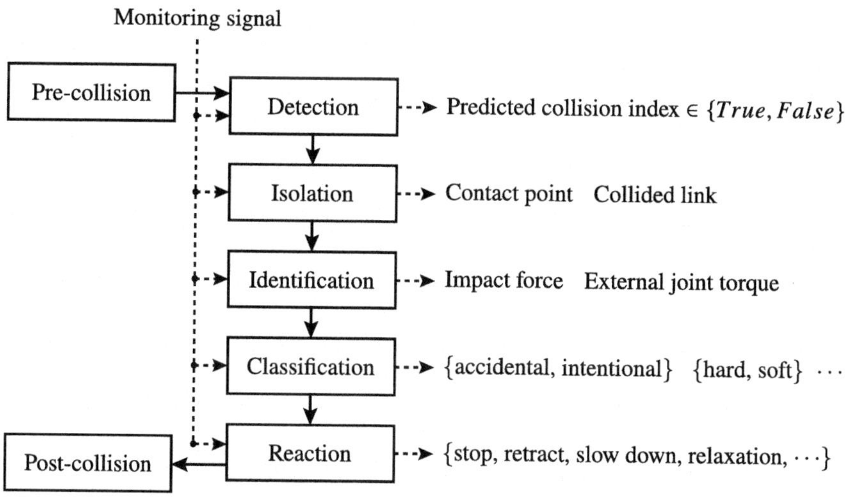

Fig. 1.2 Collision event pipeline presented in [24]

framework where collisions or contacts can be quickly and accurately detected, and case-by-case actions executed based on the nature of the collision, are essential for meaningful pHRI.

In [24] a collision handling framework is presented, in which the initial detection of the collision to the final action of the robot are organized into seven phases called the *collision event pipeline* (see Fig. 1.2). The first and last phases of the collision event pipeline are the pre-collision and post-collision phases. The pre-collision phase includes collision avoidance, while in the post-collision phase the robot decides how to proceed after the five intermediate phases have been completed. The five intermediate phases are, in order, detection, isolation, identification, classification, and reaction.

To achieve the objective for each phase, proper usage of *monitoring signals*, usually in conjunction with available kinematics or dynamic models of the robot, is essential. Monitoring signals provide information about both the collision and status of the robot, and can include signals such as joint positions and velocities, motor currents, and joint torques, as well as other information from additional sensors like tactile surface sensors.

The objectives of the five main phases are as follows:

- **Detection**: Fast and reliable detection of collisions is crucial, since if a collision is not detected at the initial phase, the subsequent phases become meaningless. The output of the detection phase is a binary index that indicates whether or not a collision has occurred (*True* for collision and *False* for no collision). One important factor that has to be considered here is the level of *detection sensitivity*. A highly sensitive detector allows for faster detection and a lower rate of false negatives (a false negative corresponds to when a collision that actually occurs is not detected), but may also lead to a higher rate of false positives (or false alarms), especially in the presence of noise in the monitoring signals. Conversely, if the detection sensitivity is too low, the risk of detection failure commensurately increases (that is, the rate of false negatives increases). Choosing the optimal level of detection sensitivity for a given range of tasks naturally has important practical ramifications.

- **Isolation**: The goal of this phase is to identify the links at which collisions have occurred (more specifically, to precisely localize the contact points of the collisions). Tactile surface sensors can be utilized to directly identify the contact point. Methods that require joint signals together with a robot model, or learning-based methods have also been developed [25–27].
- **Identification**: In this phase, the impact force of a collision, or the external joint torque resulting from the collision, is estimated [28–30]. Information provided from this phase can include not only the intensity of the impact but also the direction of the force, which can be useful for determining the reaction of the robot.
- **Classification**: Classifying the detected contact according to the information provided from the previous phases is also important for determining the proper robot reaction. For example, as described earlier, the human operator may intentionally touch (e.g., grab, push, or pull) the robot for direct pHRI. In this case, determining whether the contact is accidental or intentional is required for satisfactory collaboration. Recently, learning techniques have gained popularity as a means of classifying the contacts [31–34].
- **Reaction**: Using information gathered from the previous phases, the robot must now react to the detected collision. The simplest action the robot can take is to immediately stop, which can be a good option if the human operator can freely retract from the dangerous region after the robot has stopped. However, the operator can be placed between the robot and another fixed object, resulting in a dangerous situation. Other reaction strategies such as safely retracting from the contact, or switching to a compliant mode can be used.

1.3 Collision Detection for Robot Manipulators

In the previous sections, we have tried to make the case that collision detection is the first and the most important step to ensuring safety. We now provide a brief summary of existing collision detection methods for robot manipulators; further details are provided in later chapters.

Existing (classical) collision detection methods can be loosely divided into two categories depending on whether the dynamics model of the robot is used (model-based methods) or not (model-free methods). Model-free methods typically rely on additional external sensors attached to the manipulator, such as tactile surface sensors [35–41], three-axis accelerometers [42–45], and microphones [46, 47], or placed around the robot, such as RGB-D cameras [48, 49]. However, from both a cost and performance perspective, model-based methods that rely on a robot dynamics model together with joint torque (motor current) measurements have been more successful than model-free methods. Basic model-based methods include the energy observer-based estimation of external power [50] and estimation of external joint torques using direct estimation [24], the velocity observer [51], and the momentum observer [50, 52]. Among the four methods, the general consensus is that the momentum observer-based method is the most accurate, reliable, and cost-effective, with several works that propose improvements to the performance of the momentum observer [53–55].

The external joint torque estimated using model-based methods is known to be highly sensitive to frictional torques in the joints. To improve detection performance, several approaches that address dissipative frictional torque have been developed. Attempts have been made to accurately model and identify the friction parameters [56–59]. Because friction is highly nonlinear, discontinuous in some regimes, exhibits complex behavior (especially during velocity reversal), and also changes with the payload of the robot [60], it is extremely difficult to find an accurate friction model valid over a range of operating regimes.

To address the modeling challenges of friction, recent works have proposed learning-based methods that automatically model and learn the frictional torque [61, 62]. Methods have been proposed that circumvent friction modeling when additional sensors are available, e.g., a force-torque sensor installed at the robot's base [63], or joint torque sensors attached to the joints [24]. Other alternative model-based methods include compensation methods for uncertainties in the dynamics model [64, 65], a direct estimation-based method that requires IMU sensors attached to the links [66], and a set of techniques that can be used for the special case of robots performing cyclic (repetitive) motions [67, 68].

Recently, learning-based methods for collision detection have gained popularity as a means of sidestepping some of the implementation difficulties of pure model-based methods and compensating for uncertain dynamic effects. A learning algorithm is trained with a set of diverse, good-quality robot collision data. Based on the learning algorithm used, existing learning-based detection methods can be divided into two types: those using supervised learning algorithms and those that rely on unsupervised anomaly detection algorithms (unsupervised learning). Supervised learning-based methods typically require the true labels of the training data, which include binary true collision index and torque values, with most methods using both the collision and collision-free data for training. One approach is to use neural networks to estimate the external joint torque without the full dynamics model [69, 70]. Another set of methods learns the true collision index to either directly output a binary predicted collision index [67, 71–73] or a scalar collision prediction score [72].

Although supervised learning-based methods often perform at a satisfactory level, there are two major practical limitations to these methods: (i) Collection of robot collision data is required, and quite often the robot may become damaged as a result of repeated collisions. It is moreover impractical to collect collision data for each mass-produced industrial robots; (ii) As with supervised learning methods in general, only scenarios that are learned as collisions can be robustly detected; for accurate detection, every possible type of collision that can occur must be learned.

To address these limitations of supervised learning-based methods for collision detection, and also to enhance the practicality of collision learning (for example, making them easier to be replicated for mass-produced industrial robots), methods that use unsupervised anomaly detection algorithms [74, 75] and transfer learning-based methods [73] have been proposed.

1.4 Objectives

This book attempts to provide a concise, reasonably comprehensive (but not exhaustive) survey of collision detection methods that have been developed for robot manipulators. Beginning with a review of robot models and preliminaries on basic statistical learning methods, the book discusses the fundamentals of the collision detection problem including collision types and collision detection performance criteria, model-free versus model-based methods, and the more recent data-driven learning-based approaches to collision detection.

Special effort has been given to describing and evaluating existing methods with a unified set of notation, systematically categorizing these methods according to a basic set of criteria, and summarizing the advantages and disadvantages of each method. As far as we know, this book is the first to comprehensively organize the growing body of learning-based collision detection methods, ranging from basic supervised learning-based methods to more advanced approaches based on unsupervised learning and transfer learning techniques. For some of the learning-based methods that we have developed, step-by-step implementation details and pseudocode descriptions of the algorithms are provided. For collision detection performance criteria, we cover not only conventional criteria such as detection delay and the number of false alarms but also learning-related criteria such as generalization capability.

We hope the contents of this book will be useful to robotics engineers interested in endowing their robots with collision detection capabilities, particularly the algorithmic and software aspects of learning-based approaches. Whether it be for both research or commercial applications, or in settings ranging from industrial factories to laboratories conducting experiments on complex physical human-robot interaction tasks, we believe the book can help the reader choose and successfully implement the most appropriate detection method that suits their robot system and application.

1.5 Outline

Chapter 2 describes the fundamentals of collision detection problem handling and briefly reviews the preliminaries that may help the reader better understand the detection methods to be introduced. First, the dynamics model of rigid joint robots and flexible joint robots are reviewed. Then, the types of collision, detection sensitivity adjustment, and collision detection performance criteria are described. Finally, the learning algorithms related to the learning-based detection methods we describe in detail in Chaps. 4 and 5 are reviewed, which include support vector machine regression (SVMR), one-dimensional convolutional neural network (1-D CNN), unsupervised anomaly detection, one-class support vector machine (OC-SVM), and autoencoder-based detection.

Chapter 3 introduces various (classical) collision detection methods for robot manipulators. First, some model-free methods that use additional exteroceptive sensors are introduced. Then, we describe basic model-based methods together with pseudocode descriptions of the algorithms for real system implementation. Also, methods to deal with the frictional torque in the joints and other alternative model-based methods that improve detection performance are reviewed.

Chapter 4 reviews existing supervised learning-based approaches, which can be categorized into two groups: one that requires data for both collision and collision-free motions for training and the other that requires only the data for collision-free motions. Then, we describe in detail the two supervised learning-based methods we have developed—one based on SVMR, and the other using a 1-D CNN. After presenting the robot collision dataset we have collected, we describe in detail for each method the procedure for input feature design, the training methodology of the learning algorithms, and the collision detection (inference) methodology using the trained learning algorithms. The detection performance of the methods is also verified.

Chapter 5 introduces several approaches made to overcome the limitations of supervised learning-based detection methods, including methods that use unsupervised anomaly detection algorithms and a transfer learning-based scheme. In addition, the categorization of learning-based detection methods according to several criteria is covered in this chapter. Then, we describe in detail the two unsupervised learning-based detection methods we have developed, one using an OC-SVM and the other based on an autoencoder, together with detection performance analysis.

Chapter 6 concludes the book with a summary and discussion of the presented contents and possible future works.

References

1. Vysocky, A., Novak, P.: Human-robot collaboration in industry. MM Sci. J. **9**(2), 903–906 (2016)
2. Robla-Gómez, S., Becerra, V.M., Llata, J.R., Gonzalez-Sarabia, E., Torre-Ferrero, C., Perez-Oria, J.: Working together: a review on safe human-robot collaboration in industrial environments. IEEE Access **5**, 26754–26773 (2017)
3. Villani, V., Pini, F., Leali, F., Secchi, C.: Survey on human-robot collaboration in industrial settings: safety, intuitive interfaces and applications. Mechatronics **55**, 248–266 (2018)
4. Vicentini, F.: Collaborative robotics: a survey. J. Mech. Des. **143**(4) (2021)
5. ISO/TS 15066:2016: Robots and Robotic Devices—Collaborative Robots. International Organization for Standardization (2016)
6. Yamada, Y., Morizono, T., Umetani, Y., Takahashi, H.: Highly soft viscoelastic robot skin with a contact object-location-sensing capability. IEEE Trans. Ind. Electron. **52**(4), 960–968 (2005)

7. Kim, J., Alspach, A., Yamane, K.: 3D printed soft skin for safe human-robot interaction. In: 2015 IEEE/RSJ International Conference on Intelligent Robots and Systems (IROS), pp. 2419–2425. IEEE (2015)
8. Kim, T., Yoon, S.J., Park, Y.-L.: Soft inflatable sensing modules for safe and interactive robots. IEEE Rob. Autom. Lett. **3**(4), 3216–3223 (2018)
9. Kashiri, N., Laffranchi, M., Tsagarakis, N.G., Margan, A., Caldwell, D.G.: Physical interaction detection and control of compliant manipulators equipped with friction clutches. In: 2014 IEEE International Conference on Robotics and Automation (ICRA), pp. 1066–1071. IEEE (2014)
10. Park, S., Park, K., Jeong, H., Shin, W., Kim, J.: A safe and rapidly switchable stiffness hydrostatic actuator through valve-controlled air springs. In: 2021 IEEE/RSJ International Conference on Intelligent Robots and Systems (IROS), pp. 9333–9338. IEEE (2021)
11. Niu, Z., Awad, M.I., Shah, U.H., Boushaki, M.N., Zweiri, Y., Seneviratne, L., Hussain, I.: Towards safe physical human-robot interaction by exploring the rapid stiffness switching feature of discrete variable stiffness actuation. IEEE Rob. Autom. Lett. **7**(3), 8084–8091 (2022)
12. Albu-Schäffer, A., Haddadin, S., Ott, C., Stemmer, A., Wimböck, T., Hirzinger, G.: The DLR lightweight robot: design and control concepts for robots in human environments. Ind. Rob. Int. J. **34**(5), 376–385 (2007)
13. Bischoff, R., Kurth, J., Schreiber, G., Koeppe, R., Albu-Schäffer, A., Beyer, A., Eiberger, O., Haddadin, S., Stemmer, A., Grunwald, G., et al.: The KUKA-DLR lightweight robot arm—a new reference platform for robotics research and manufacturing. In: ISR 2010 (41st International Symposium on Robotics) and ROBOTIK 2010 (6th German Conference on Robotics), pp. 1–8. VDE (2010)
14. Haddadin, S., Parusel, S., Johannsmeier, L., Golz, S., Gabl, S., Walch, F., Sabaghian, M., Jähne, C., Hausperger, L., Haddadin, S.: The Franka Emika robot: a reference platform for robotics research and education. IEEE Rob. Autom. Mag. **29**(2), 46–64 (2022)
15. Diankov, R., Kuffner, J.: OpenRAVE: a planning architecture for autonomous robotics. Robotics Institute, Pittsburgh, PA, Technical Report CMU-RI-TR-08-34, 79 (2008)
16. Sisbot, E.A., Alami, R.: A human-aware manipulation planner. IEEE Trans. Rob. **28**(5), 1045–1057 (2012)
17. Mainprice, J., Berenson, D.: Human-robot collaborative manipulation planning using early prediction of human motion. In: 2013 IEEE/RSJ International Conference on Intelligent Robots and Systems, pp. 299–306. IEEE (2013)
18. Park, S.-O., Lee, M.C., Kim, J.: Trajectory planning with collision avoidance for redundant robots using Jacobian and artificial potential field-based real-time inverse kinematics. Int. J. Control Autom. Syst. **18**(8), 2095–2107 (2020)
19. Yu, Y.-H., Zhang, Y.-T.: Collision avoidance and path planning for industrial manipulator using slice-based heuristic fast marching tree. Rob. Comput. Integr. Manuf. **75**, 102289 (2022)
20. Flacco, F., Kröger, T., De Luca, A., Khatib, O.: A depth space approach to human-robot collision avoidance. In: 2012 IEEE International Conference on Robotics and Automation, pp. 338–345. IEEE (2012)
21. Mohammed, A., Schmidt, B., Wang, L.: Active collision avoidance for human-robot collaboration driven by vision sensors. Int. J. Comput. Integr. Manuf. **30**(9), 970–980 (2017)
22. Safeea, M., Neto, P., Bearee, R.: On-line collision avoidance for collaborative robot manipulators by adjusting off-line generated paths: an industrial use case. Rob. Auton. Syst. **119**, 278–288 (2019)
23. Liu, G., He, H., Tian, G., Zhang, J., Ji, Z.: Online collision avoidance for human-robot collaborative interaction concerning safety and efficiency. In: 2020 IEEE/ASME International Conference on Advanced Intelligent Mechatronics (AIM), pp. 1667–1672. IEEE (2020)
24. Haddadin, S., De Luca, A., Albu-Schäffer, A.: Robot collisions: a survey on detection, isolation, and identification. IEEE Trans. Rob. **33**(6), 1292–1312 (2017)
25. Manuelli, L., Tedrake, R.: Localizing external contact using proprioceptive sensors: the contact particle filter. In: 2016 IEEE/RSJ International Conference on Intelligent Robots and Systems (IROS), pp. 5062–5069. IEEE (2016)

26. Briquet-Kerestedjian, N., Wahrburg, A., Grossard, M., Makarov, M., Rodriguez-Ayerbe, P.: Using neural networks for classifying human-robot contact situations. In: 2019 18th European Control Conference (ECC), pp. 3279–3285. IEEE (2019)
27. Liang, J., Kroemer, O.: Contact localization for robot arms in motion without torque sensing. In: 2021 IEEE International Conference on Robotics and Automation (ICRA), pp. 6322–6328. IEEE (2021)
28. Wahrburg, A., Morara, E., Cesari, G., Matthias, B., Ding, H.: Cartesian contact force estimation for robotic manipulators using Kalman filters and the generalized momentum. In: 2015 IEEE International Conference on Automation Science and Engineering (CASE), pp. 1230–1235. IEEE (2015)
29. Hu, J., Xiong, R.: Contact force estimation for robot manipulator using semiparametric model and disturbance Kalman filter. IEEE Trans. Ind. Electron. **65**(4), 3365–3375 (2017)
30. Wahrburg, A., Bös, J., Listmann, K.D., Dai, F., Matthias, B., Ding, H.: Motor-current-based estimation of Cartesian contact forces and torques for robotic manipulators and its application to force control. IEEE Trans. Autom. Sci. Eng. **15**(2), 879–886 (2017)
31. Golz, S., Osendorfer, C., Haddadin, S.: Using tactile sensation for learning contact knowledge: discriminate collision from physical interaction. In: 2015 IEEE International Conference on Robotics and Automation (ICRA), pp. 3788–3794. IEEE (2015)
32. Anvaripour, M., Saif, M.: Collision detection for human-robot interaction in an industrial setting using force myography and a deep learning approach. In: 2019 IEEE International Conference on Systems, Man and Cybernetics (SMC), pp. 2149–2154. IEEE (2019)
33. Cioffi, G., Klose, S., Wahrburg, A.: Data-efficient online classification of human-robot contact situations. In: 2020 European Control Conference (ECC), pp. 608–614. IEEE (2020)
34. Lippi, M., Gillini, G., Marino, A., Arrichiello, F.: A data-driven approach for contact detection, classification and reaction in physical human-robot collaboration. In: 2021 IEEE International Conference on Robotics and Automation (ICRA), pp. 3597–3603. IEEE (2021)
35. Duchaine, V., Lauzier, N., Baril, M., Lacasse, M.-A., Gosselin, C.: A flexible robot skin for safe physical human robot interaction. In: 2009 IEEE International Conference on Robotics and Automation, pp. 3676–3681. IEEE (2009)
36. Strohmayr, M.W., Wörn, H., Hirzinger, G.: The DLR artificial skin step I: uniting sensitivity and collision tolerance. In: 2013 IEEE International Conference on Robotics and Automation, pp. 1012–1018. IEEE (2013)
37. Cirillo, A., Ficuciello, F., Natale, C., Pirozzi, S., Villani, L.: A conformable force/tactile skin for physical human-robot interaction. IEEE Rob. Autom. Lett. **1**(1), 41–48 (2015)
38. Hughes, D., Lammie, J., Correll, N.: A robotic skin for collision avoidance and affective touch recognition. IEEE Rob. Autom. Lett. **3**(3), 1386–1393 (2018)
39. Pang, G., Yang, G., Heng, W., Ye, Z., Huang, X., Yang, H.-Y., Pang, Z.: CoboSkin: soft robot skin with variable stiffness for safer human-robot collaboration. IEEE Trans. Ind. Electron. **68**(4), 3303–3314 (2020)
40. Ye, Z., Pang, G., Xu, K., Hou, Z., Lv, H., Shen, Y., Yang, G.: Soft robot skin with conformal adaptability for on-body tactile perception of collaborative robots. IEEE Rob. Autom. Lett. **7**(2), 5127–5134 (2022)
41. Ji, S., Shin, J., Yoon, J., Lim, K.-H., Sim, G.-D., Lee, Y.-S., Kim, D.H., Cho, H., Park, J.: Three-dimensional skin-type triboelectric nanogenerator for detection of two-axis robotic-arm collision. Nano Energy **97**, 107225 (2022)
42. Wisanuvej, P., Liu, J., Chen, C.-M., Yang, G.-Z.: Blind collision detection and obstacle characterisation using a compliant robotic arm. In: 2014 IEEE International Conference on Robotics and Automation (ICRA), pp. 2249–2254. IEEE (2014)
43. Min, F., Wang, G., Liu, N.: Collision detection and identification on robot manipulators based on vibration analysis. Sensors **19**(5), 1080 (2019)
44. Sandykbayeva, D., Kappassov, Z., Orazbayev, B.: VibroTouch: active tactile sensor for contact detection and force sensing via vibrations. Sensors **22**(17), 6456 (2022)
45. Valle, C.M.C.O., Kurdas, A., Fortunić, E.P., Abdolshah, S., Haddadin, S.: Real-time IMU-based learning: a classification of contact materials. In: 2022 IEEE/RSJ International Conference on Intelligent Robots and Systems (IROS), pp. 1965–1971. IEEE (2022)

46. Fan, X., Lee, D., Chen, Y., Prepscius, C., Isler, V., Jackel, L., Sebastian Seung, H., Lee, D.: Acoustic collision detection and localization for robot manipulators. In: 2020 IEEE/RSJ International Conference on Intelligent Robots and Systems (IROS), pp. 9529–9536. IEEE (2020)
47. Dimiccoli, M., Patni, S., Hoffmann, M., Moreno-Noguer, F.: Recognizing object surface material from impact sounds for robot manipulation. In: 2022 IEEE/RSJ International Conference on Intelligent Robots and Systems (IROS), pp. 9280–9287. IEEE (2022)
48. Maric, B., Jurican, F., Orsag, M., Kovacic, Z.: Vision based collision detection for a safe collaborative industrial manipulator. In: 2021 IEEE International Conference on Intelligence and Safety for Robotics (ISR), pp. 334–337. IEEE (2021)
49. Makris, S., Aivaliotis, P.: AI-based vision system for collision detection in HRC applications. Procedia CIRP **106**, 156–161 (2022)
50. De Luca, A., Albu-Schaffer, A., Haddadin, S., Hirzinger, G.: Collision detection and safe reaction with the DLR-III lightweight manipulator arm. In: 2006 IEEE/RSJ International Conference on Intelligent Robots and Systems, pp. 1623–1630. IEEE (2006)
51. Haddadin, S.: Towards Safe Robots: Approaching Asimov's 1st Law, vol. 90. Springer, Berlin (2013)
52. De Luca, A., Mattone, R.: Sensorless robot collision detection and hybrid force/motion control. In: Proceedings of the 2005 IEEE International Conference on Robotics and Automation, pp. 999–1004. IEEE (2005)
53. Garofalo, G., Mansfeld, N., Jankowski, J., Ott, C.: Sliding mode momentum observers for estimation of external torques and joint acceleration. In: 2019 International Conference on Robotics and Automation (ICRA), pp. 6117–6123. IEEE (2019)
54. Li, Y., Li, Y., Zhu, M., Xu, Z., Mu, D.: A nonlinear momentum observer for sensorless robot collision detection under model uncertainties. Mechatronics **78**, 102603 (2021)
55. Huang, S., Gao, M., Liu, L., Chen, J., Zhang, J.: A back-input compensation approach. IEEE/ASME Trans. Mechatron. Collision Detect. Cobots (2022)
56. Lee, S.-D., Song, J.-B.: Sensorless collision detection based on friction model for a robot manipulator. Int. J. Precis. Eng. Manuf. **17**(1), 11–17 (2016)
57. Gaz, C., Magrini, E., De Luca, A.: A model-based residual approach for human-robot collaboration during manual polishing operations. Mechatronics **55**, 234–247 (2018)
58. Zhang, C., Mu, C., Wang, Y., Li, J., Liu, Z.: Collision detection for six-DOF serial robots force/position hybrid control based on continuous friction model. Meas. Control 00202940221091575 (2022)
59. Xiao, J., Zhang, Q., Hong, Y., Wang, G., Zeng, F.: Collision detection algorithm for collaborative robots considering joint friction. Int. J. Adv. Rob. Syst. **15**(4), 1729881418788992 (2018)
60. Swevers, J., Verdonck, W., De Schutter, J.: Dynamic model identification for industrial robots. IEEE Control Syst. Mag. **27**(5), 58–71 (2007)
61. Xu, T., Fan, J., Fang, Q., Zhu, Y., Zhao, J.: A new robot collision detection method: a modified nonlinear disturbance observer based-on neural networks. J. Intell. Fuzzy Syst. **38**(1), 175–186 (2020)
62. Lim, D., Kim, D., Park, J.: Momentum observer-based collision detection using LSTM for model uncertainty learning. In: 2021 IEEE International Conference on Robotics and Automation (ICRA), pp. 4516–4522. IEEE (2021)
63. Li, W., Han, Y., Wu, J., Xiong, Z.: Collision detection of robots based on a force/torque sensor at the bedplate. IEEE/ASME Trans. Mechatron. **25**(5), 2565–2573 (2020)
64. Makarov, M., Caldas, A., Grossard, M., Rodriguez-Ayerbe, P., Dumur, D.: Adaptive filtering for robust proprioceptive robot impact detection under model uncertainties. IEEE/ASME Trans. Mechatron. **19**(6), 1917–1928 (2014)
65. Birjandi, S.A.B., Haddadin, S.: Model-adaptive high-speed collision detection for serial-chain robot manipulators. IEEE Rob. Autom. Lett. **5**(4), 6544–6551 (2020)
66. Birjandi, S.A.B., Kühn, J., Haddadin, S.: Observer-extended direct method for collision monitoring in robot manipulators using proprioception and IMU sensing. IEEE Rob. Autom. Lett. **5**(2), 954–961 (2020)

67. Heo, Y.J., Kim, D., Lee, W., Kim, H., Park, J., Chung, W.K.: Collision detection for industrial collaborative robots: a deep learning approach. IEEE Rob. Autom. Lett. **4**(2), 740–746 (2019)
68. Gordić, Z., Jovanović, K.: Collision detection on industrial robots in repetitive tasks using modified dynamic time warping. Robotica **38**(10), 1717–1736 (2020)
69. Sharkawy, A.-N., Koustoumpardis, P.N., Aspragathos, N.: Neural network design for manipulator collision detection based only on the joint position sensors. Robotica **38**(10), 1737–1755 (2020)
70. Sharkawy, A.-N., Koustoumpardis, P.N., Aspragathos, N.: Human–robot collisions detection for safe human–robot interaction using one multi-input–output neural network. Soft Comput. **24**(9), 6687–6719 (2020)
71. Zhang, Z., Qian, K., Schuller, B.W., Wollherr, D.: An online robot collision detection and identification scheme by supervised learning and Bayesian decision theory. IEEE Trans. Autom. Sci. Eng. (2020)
72. Park, K.M., Kim, J., Park, J., Park, F.C.: Learning-based real-time detection of robot collisions without joint torque sensors. IEEE Rob. Autom. Lett. **6**(1), 103–110 (2020)
73. Kim, D., Lim, D., Park, J.: Transferable collision detection learning for collaborative manipulator using versatile modularized neural network. IEEE Trans. Rob. (2021)
74. Chen, T., Liu, X., Xia, B., Wang, W., Lai, Y.: Unsupervised anomaly detection of industrial robots using sliding-window convolutional variational autoencoder. IEEE Access **8**, 47072–47081 (2020)
75. Park, K.M., Park, Y., Yoon, S., Park, F.C.: Collision detection for robot manipulators using unsupervised anomaly detection algorithms. IEEE/ASME Trans. Mechatron. (2021)

Chapter 2
Fundamentals

2.1 Introduction

In this chapter, we first briefly review the dynamics model of robot manipulators, for both rigid joint and flexible joint robots. Then, we describe the fundamentals of collision detection, which are very important when handling the collision detection problem but have not been organized for easy identification in the existing literature. The above fundamentals include types of collision, detection sensitivity adjustment, and collision detection performance criteria. Lastly, the learning algorithms related to the learning-based detection methods we describe in detail in Chaps. 4 and 5 are reviewed, which include SVMR, 1-D CNN, the main concept of unsupervised anomaly detection, OC-SVM, and autoencoder-based detection.

2.2 Robot Dynamics

2.2.1 Rigid Joint Robots

In this section, we review the dynamics of an n-DOF open-chain rigid joint robot manipulator (Fig. 2.1 shows an illustration of a 6-DOF open-chain robot manipulator). The generalized coordinates are the link-side joint position $q \in \mathbb{R}^n$, and the motor-side joint position $\theta \in \mathbb{R}^n$ is assumed to coincide with q. The dynamics of the robot can be expressed as

$$M(q)\ddot{q} + C(q, \dot{q})\dot{q} + g(q) + \tau_F = \tau_m, \tag{2.1}$$

where $M(q) \in \mathbb{R}^{n\times n}$ is the symmetric and positive-definite mass matrix, $C(q, \dot{q}) \in \mathbb{R}^{n\times n}$ is the Coriolis matrix, $g(q) \in \mathbb{R}^n$ is the gravity vector, $\tau_F \in \mathbb{R}^n$ is the torque attributable to dissipative friction acting on the motor side of the joints, and $\tau_m \in \mathbb{R}^n$

K. M. Park and F. C. Park, *Collision Detection for Robot Manipulators: Methods and Algorithms*, Springer Tracts in Advanced Robotics 155,
https://doi.org/10.1007/978-3-031-30195-7_2

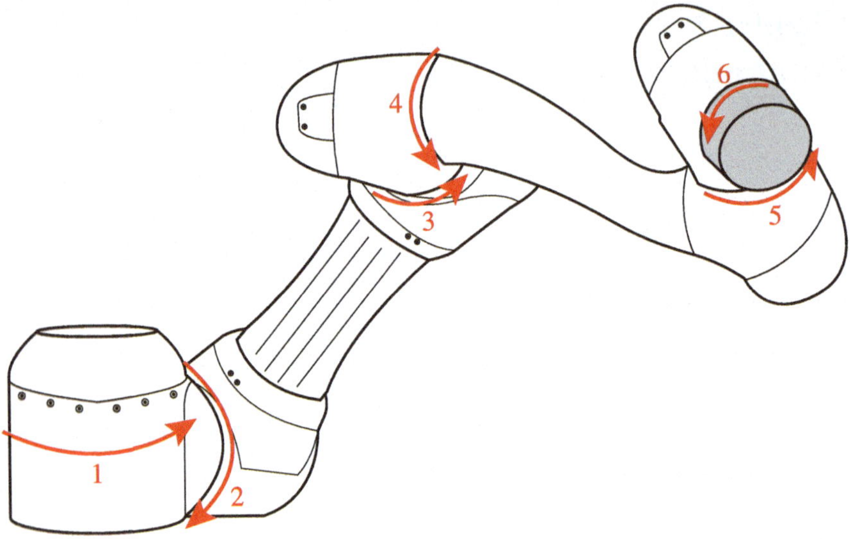

Fig. 2.1 A 6-DOF open-chain robot manipulator

is the active motor torque which can be calculated as $\tau_m = K_i i_m$, where $K_i \in \mathbb{R}^{n \times n}$ is the positive-definite diagonal current-to-torque gain matrix and $i_m \in \mathbb{R}^n$ is the motor current. When a collision or other external force is applied to the robot, the dynamics model of the robot (2.1) becomes

$$M(q)\ddot{q} + C(q, \dot{q})\dot{q} + g(q) + \tau_F = \tau_m + \tau_{ext}, \tag{2.2}$$

where $\tau_{ext} \in \mathbb{R}^n$ is the external joint torque due to the collision or other external force applied to the robot.

The passivity property of the robot dynamics implies that the matrix $\dot{M}(q) - 2C(q, \dot{q})$ is skew-symmetric, where $\dot{M}(q)$ is the time derivative of $M(q)$ [1]. The property can be expressed as

$$\begin{aligned} \dot{M}(q) - 2C(q, \dot{q}) &= -(\dot{M}(q) - 2C(q, \dot{q}))^T \\ &= -\dot{M}(q) + 2C^T(q, \dot{q}), \end{aligned} \tag{2.3}$$

and therefore is equivalent to the identity

$$\dot{M}(q) = C(q, \dot{q}) + C^T(q, \dot{q}). \tag{2.4}$$

We first introduce the robot's generalized momentum $p \in \mathbb{R}^n$, which is defined as

$$p = M(q)\dot{q}. \tag{2.5}$$

The time derivative of p is expressed as

$$\begin{aligned}\dot{p} &= \dot{M}(q)\dot{q} + M(q)\ddot{q} \\ &= (C(q,\dot{q}) + C^T(q,\dot{q}))\dot{q} + \tau_m + \tau_{ext} - \tau_F - C(q,\dot{q})\dot{q} - g(q) \\ &= \tau_m + \tau_{ext} - \tau_F + C^T(q,\dot{q})\dot{q} - g(q),\end{aligned} \tag{2.6}$$

where the robot dynamics with external joint torque (2.2) and the robot dynamics identity (2.4) are used.

Next, the robot's total energy $E \in \mathbb{R}$ is defined as

$$\begin{aligned}E &= T(q,\dot{q}) + U(q) \\ &= \frac{1}{2}\dot{q}^T M(q)\dot{q} + U(q),\end{aligned} \tag{2.7}$$

where $T(q,\dot{q}) \in \mathbb{R}$ is the kinetic energy of the robot and $U(q) \in \mathbb{R}$ is the potential energy due to gravity with $g(q) = (\partial U/\partial q)^T$. The time derivative of E is expressed as

$$\begin{aligned}\dot{E} &= \frac{1}{2}\ddot{q}^T M(q)\dot{q} + \frac{1}{2}\dot{q}^T \dot{M}(q)\dot{q} + \frac{1}{2}\dot{q}^T M(q)\ddot{q} + \dot{U}(q) \\ &= \dot{q}^T M(q)\ddot{q} + \frac{1}{2}\dot{q}^T \dot{M}(q)\dot{q} + \dot{q}^T g(q) \\ &= \dot{q}^T(\tau_m + \tau_{ext} - \tau_F - C(q,\dot{q})\dot{q} - g(q)) + \frac{1}{2}\dot{q}^T \dot{M}(q)\dot{q} + \dot{q}^T g(q) \\ &= \dot{q}^T\tau_m + \dot{q}^T\tau_{ext} - \dot{q}^T\tau_F + \frac{1}{2}\dot{q}^T(\dot{M}(q) - 2C(q,\dot{q}))\dot{q} \\ &= \dot{q}^T\tau_m + \dot{q}^T\tau_{ext} - \dot{q}^T\tau_F,\end{aligned} \tag{2.8}$$

where the robot dynamics with external joint torque (2.2) and the passivity property of the robot dynamics (skew-symmetry of $\dot{M}(q) - 2C(q,\dot{q})$) are used. The term $\dot{q}^T\tau_{ext}$ in (2.8) is the external power P_{ext}.

Finally, assuming only one contact (collision) point, the collision twist $\mathcal{V}_c \in \mathbb{R}^6$ and the collision wrench $\mathcal{F}_c \in \mathbb{R}^6$ are defined as

$$\mathcal{V}_c = \begin{bmatrix}\omega_c \\ v_c\end{bmatrix} = J_c(q)\dot{q}, \tag{2.9}$$

$$\mathcal{F}_c = \begin{bmatrix}m_{ext} \\ f_{ext}\end{bmatrix}, \tag{2.10}$$

where $\omega_c \in \mathbb{R}^3$ is the angular velocity of the collided robot link, $v_c \in \mathbb{R}^3$ is the linear velocity of the contact point, $m_{ext} \in \mathbb{R}^3$ and $f_{ext} \in \mathbb{R}^3$ are the moment and force due to the collision, and $J_c(q) \in \mathbb{R}^{6\times 6}$ is the contact Jacobian. The external joint torque τ_{ext} can be expressed as

$$\tau_{ext} = J_c^T(q)\mathcal{F}_c, \tag{2.11}$$

and the external power P_{ext} can be written as

$$P_{ext} = \dot{q}^T \tau_{ext} = \dot{q}^T J_c^T(q)\mathcal{F}_c = \mathcal{V}_c^T \mathcal{F}_c, \tag{2.12}$$

where (2.9) and (2.11) are used.

2.2.2 Flexible Joint Robots

For n-DOF open-chain flexible joint robot manipulators, both the link-side joint position q and the motor-side joint position θ are the generalized coordinates, where the joint deflection defined as $\delta = \theta - q$ cannot be neglected. Assuming no inertial coupling between the motors and the links, we take into account the reduced model of Spong [2]. The link dynamics of the robot is expressed as

$$M(q)\ddot{q} + C(q, \dot{q})\dot{q} + g(q) = \tau_J + \tau_{ext}, \tag{2.13}$$

where τ_J is the elastic joint torque due to the joint flexibility, which can be measured by joint torque sensors if available. The motor dynamics of the robot is assumed to be of the form

$$B\ddot{\theta} + \tau_J = \tau_m - \tau_F, \tag{2.14}$$

where $B \in \mathbb{R}^{n\times n}$ is the positive-definite diagonal motor inertia matrix. Unlike friction acting on the link side of the joints which can often be safely neglected, τ_F is usually not negligible. The elastic joint torque τ_J couples the link and the motor dynamics, with its dependence on δ taking on a variety of forms. We refer the reader to [3, 4] for further details.

The generalized momentum of the flexible joint robot $p_J \in \mathbb{R}^{2n}$ is defined as

$$p_J = \begin{bmatrix} p_q \\ p_\theta \end{bmatrix} = \begin{bmatrix} M(q)\dot{q} \\ B\dot{\theta} \end{bmatrix}, \tag{2.15}$$

where $p_q \in \mathbb{R}^n$ is the link-side momentum and $p_\theta \in \mathbb{R}^n$ is the motor-side momentum, together defining the total momentum $p_{tot} \in \mathbb{R}^n$ as $p_{tot} = p_q + p_\theta$. The time derivative of p_q is expressed as

$$\begin{aligned} \dot{p}_q &= \dot{M}(q)\dot{q} + M(q)\ddot{q} \\ &= (C(q, \dot{q}) + C^T(q, \dot{q}))\dot{q} + \tau_J + \tau_{ext} - C(q, \dot{q})\dot{q} - g(q) \\ &= \tau_J + \tau_{ext} + C^T(q, \dot{q})\dot{q} - g(q), \end{aligned} \tag{2.16}$$

where the link dynamics (2.13) and the robot dynamics identity (2.4) are used, and the time derivative of p_θ is

$$\dot{p}_\theta = B\ddot{\theta} = \tau_m - \tau_F - \tau_J, \tag{2.17}$$

where the motor dynamics (2.14) is used. Together, the time derivative of the total momentum is

$$\begin{aligned} \dot{p}_{tot} &= \dot{p}_q + \dot{p}_\theta \\ &= \tau_m - \tau_F + \tau_{ext} + C^T(q, \dot{q})\dot{q} - g(q). \end{aligned} \tag{2.18}$$

2.3 Fundamentals of Collision Detection

In this section, we describe the fundamentals of collision detection handling including the types of collision (hard and soft collisions), detection sensitivity adjustment, and collision detection performance criteria (detection delay, detection failure, and the number of false alarms).

2.3.1 Collision Types

Before we categorize the types of collision, we first describe the collection methodology of the robot collision dataset we use in Chaps. 4 and 5, together introducing some terms useful for defining the collision types [5]. The collision dataset is used for training and validating the learning-based collision detection methods we have developed, where data collection is performed with a 6-DOF collaborative robot manipulator (Doosan Robotics, *M0609*: rated payload 6 kg, weight 27 kg) as illustrated abstractly in Fig. 2.1.

To train and test the detection methods, the *true collision index* is required, which is a binary label that stays at one while a collision between the manipulator and an external object is occurring, and stays at zero otherwise. To acquire the true collision index, we use a force-sensitive resistor (FSR) (Interlink Electronics, *FSR 406*), in which the resistance decreases when force is applied to the sensor. A collision index labeling circuit is designed as depicted in Fig. 2.2a, where the output voltage stays constant and decreases only when a collision occurs (Fig. 2.3). The FSR part of the circuit is integrated into the hand-held labeling tool as shown in Fig. 2.2b. The relationship between the output voltage of the circuit and the actual force applied is calibrated as shown in Fig. 2.3a, and the binary true collision index is acquired by thresholding the voltage; if the output voltage is lower than 7.26 V (i.e., the applied force is larger than 3 N), the true index is labeled as collision. As illustrated in Fig. 2.3b, the time during which the true collision index stays at one is defined as the *collision duration time*. We denote the binary output of the collision detection algorithms, which is a predicted value of the true collision index, as the *predicted collision index*.

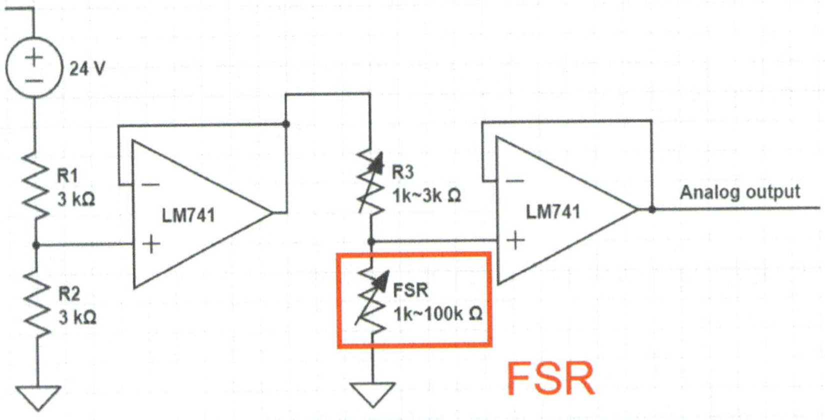

(a)

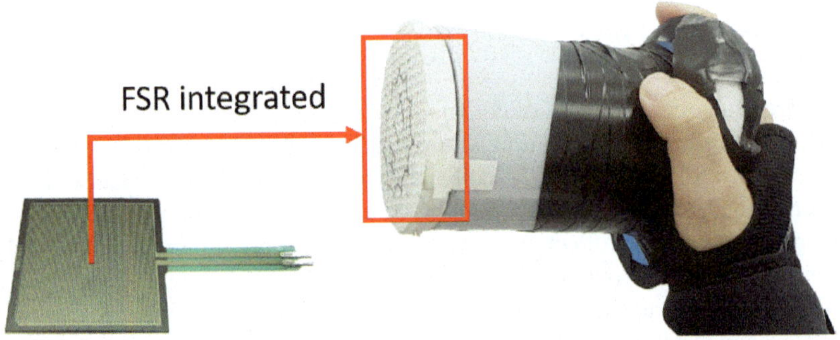

(b)

Fig. 2.2 **a** The collision index labeling circuit using an FSR. **b** The collision index labeling tool integrated with the FSR

We collect the robot collision dataset for three collision scenarios: (1) *hard* collisions (sharp impact collisions), (2) *soft* collisions (pushing and pulling motions), and (3) *collision-free* motions. Hard collisions and soft collisions have different ranges of collision duration time; hard collisions have duration time on the order of 10–100 ms and are collected by quickly striking the robot with the labeling tool, while soft collisions have duration time in the range 100–700 ms and are acquired by slowly pressing (pushing) the robot. We note that for the collision detection phase, both hard and soft collisions are treated in the same manner as just collisions, and distinguishing between the two types is addressed in the collision classification phase.

To summarize, hard collisions are transient impact collisions (e.g., striking the robot) having collision duration time on the order of 10–100 ms, while soft collisions

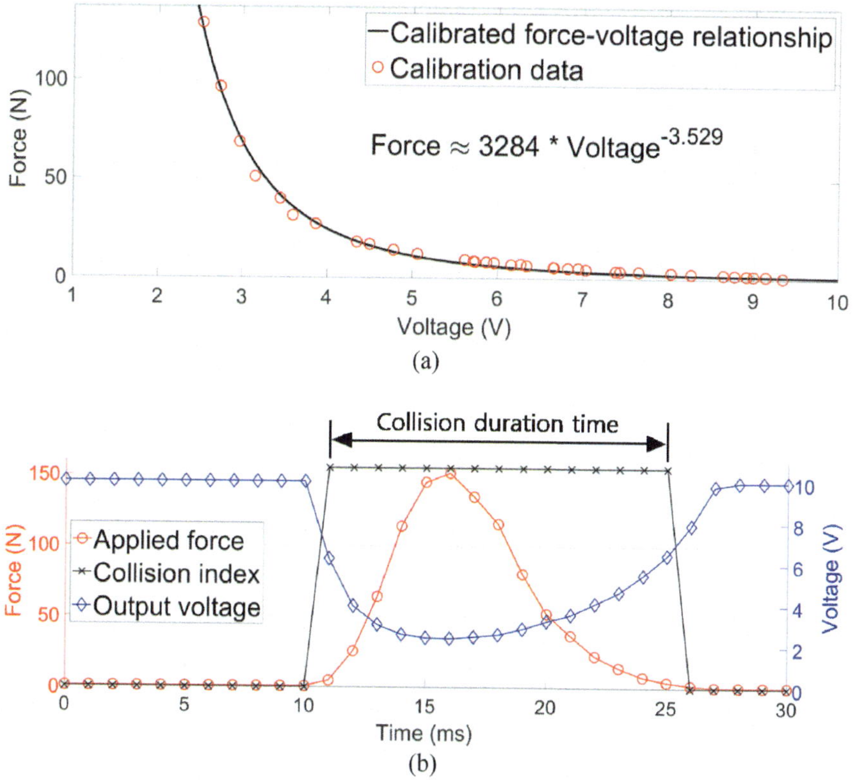

Fig. 2.3 **a** The calibrated relationship between the applied force and the circuit output voltage. **b** The output voltage, force profile, and true collision index when a collision occurs. The collision duration time is defined as shown in the figure

are quasi-static collisions (e.g., pushing, pulling, or grabbing the robot) that have a duration time in the range of 100–700 ms. Figure 2.4 shows how the collision duration times of a total of 787 test collisions (578 hard collisions and 209 soft collisions) are distributed.

2.3.2 *Detection Sensitivity Adjustment*

Most collision detection methods require detection sensitivity adjustment. For example, the momentum observer-based method, the most widely-used dynamics model-based method, estimates the vector-valued external joint torque, and the energy observer-based method, which is another model-based method, estimates the scalar-valued external power (later described in Chap. 3). Except for some end-to-end networks, most learning-based methods output either a vector or scalar of real numbers.

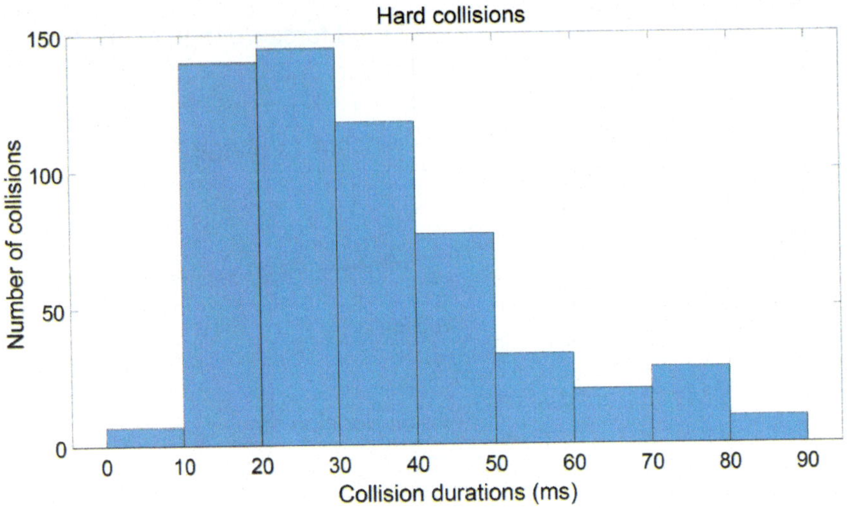

(a)

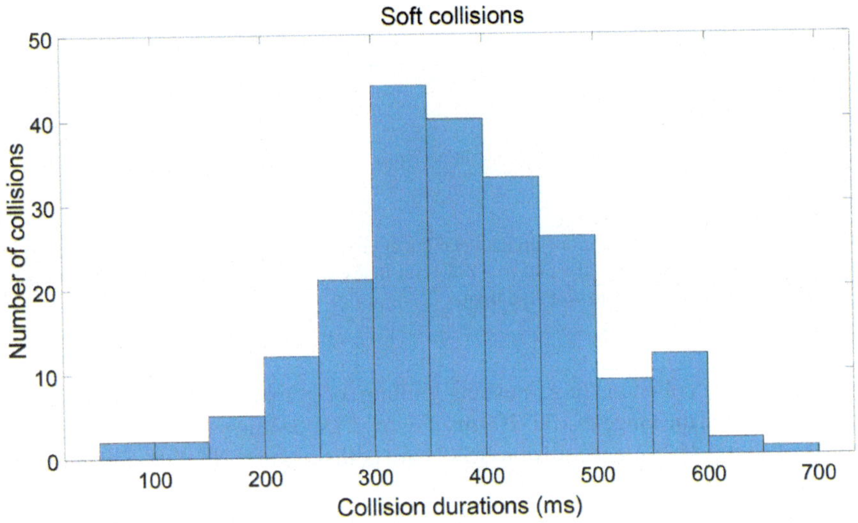

(b)

Fig. 2.4 The distribution of collision duration times of **a** 578 hard collisions and **b** 209 soft collisions

To determine whether a collision has occurred or not, what we need to know at the end is the binary predicted collision index. As we can see from the examples above, the output of most detection methods is a real number rather than a binary value, consequently requiring a thresholding process.

Assume a detection method outputs a vector output $y_v \in \mathbb{R}^m$, where $m \neq 1$ is in many cases equal to the number of joints (e.g., momentum observer-based methods, learning-based external joint torque estimation). Then, the collision decision function $cd : y_v \rightarrow \{True, False\}$ which maps the vector output y_v to the binary predicted collision index by component-wise comparing is defined usually as

$$cd(y_v) = \begin{cases} True & \text{if } |y_{v,i}| > \epsilon_{v,i} \\ False & \text{if } |y_{v,i}| \leq \epsilon_{v,i} \end{cases}, \quad i = 1, \ldots, m, \tag{2.19}$$

where $y_{v,i}$ is the ith component of y_v and $\epsilon_v \in \mathbb{R}^m$ is a user-defined detection threshold. On the other hand, if a detection method outputs a scalar value $y_s \in \mathbb{R}$ (e.g., energy observer-based methods, unsupervised learning-based anomaly score computation), the collision decision function $cd : y_s \rightarrow \{True, False\}$ is defined usually as

$$cd(y_s) = \begin{cases} True & \text{if } |y_s| > \epsilon_s \\ False & \text{if } |y_s| \leq \epsilon_s \end{cases}, \tag{2.20}$$

where y_s is compared with a scalar detection threshold $\epsilon_s \in \mathbb{R}$. Some methods use state-dependent dynamic thresholds, but here we assume that the threshold is a fixed value. We note that depending on the detection method, the absolute value computation may not be required and the direction of the inequality sign may be opposite.

Tuning the detection threshold is directly related to the sensitivity adjustment of the detection methods; high sensitivity is achieved by a low threshold and low sensitivity by a high threshold. When the threshold is set to a very small value, collisions are detected with low detection delays, reducing the risk of human injury or damage to the robot. However, due to disturbances, e.g., tremors in the robot joints, the detection algorithm may report a collision even when there is no actual collision occurrence. If the robot is programmed to immediately stop when a collision is detected, frequent stops due to false alarms will reduce productivity. Conversely, a threshold setting that is too high will result in multiple detection failures, which is dangerous for both the human and the robot. Finding the optimal threshold that ensures both safety and productivity is a critical issue.

One simple way to automatically find the threshold that is the most sensitive to collisions while not producing any false alarms is to observe the detection output $y(t)$ during collision-free motions for a sufficiently long time interval $[0, T]$ (in our case, one hour). The optimal threshold ϵ_{op} can be found by setting the threshold to the maximum or minimum value of the observed outputs, depending on the output aspect of the detection method:

$$\epsilon_{op} = \begin{cases} \max\{|y(t)|, t \in [0, T]\} \text{ or} \\ \max\{y(t), t \in [0, T]\} \text{ or} \\ \min\{y(t), t \in [0, T]\} \end{cases}, \tag{2.21}$$

Figure 2.5 shows detection results of the unsupervised learning-based methods we describe in Chap. 5 [6]. When a collision occurs, the scalar anomaly score resulting from the autoencoder-based reconstruction increases (and therefore takes the maximum value of the observed outputs as the detection threshold), while the output scalar value of the trained OC-SVM decreases (and consequently uses the minimum observation value).

Users may want to adjust the sensitivity depending on the robot usage environment, for example, if the robot is kept in a safety fence separated from humans, reducing the chance of false alarms by lowering the detection sensitivity may be a preferable option. As we can see from (2.19) and (2.20), a scalar-valued output allows more intuitive sensitivity adjustment since only one constant threshold is required to be tuned, while a vector-valued output requires manual tuning of multiple thresholds. When frequent or intuitive sensitivity tuning is needed, one can consider using a detection method that reports a scalar output.

2.3.3 *Collision Detection Performance Criteria*

2.3.3.1 Area Under the Precision-Recall Curve (PRAUC)

Precision-recall is a useful criterion for evaluating the performance of the classification (on a per-sample basis) of highly imbalanced classes, which corresponds to the collision detection problem in which there are significantly fewer collision-labeled samples than collision-free-labeled samples (in our case 95,681 samples versus 3,743,905 samples). Precision and recall are defined as

$$Precision = \frac{TP}{TP + FP} \tag{2.22}$$

$$Recall = \frac{TP}{TP + FN}, \tag{2.23}$$

where TP is the number of true positives, FP is the number of false positives, and FN is the number of false negatives (collisions are considered positives). High precision indicates a low false positive rate while high recall indicates a low false negative rate. With varying detection thresholds (only applicable to scalar-valued detection outputs), the precision-recall curve shows the trade-off between precision and recall which can be considered the *global performance* of a classifier, where a high PRAUC represents both high precision and recall. Figure 2.6 shows an example of a precision-

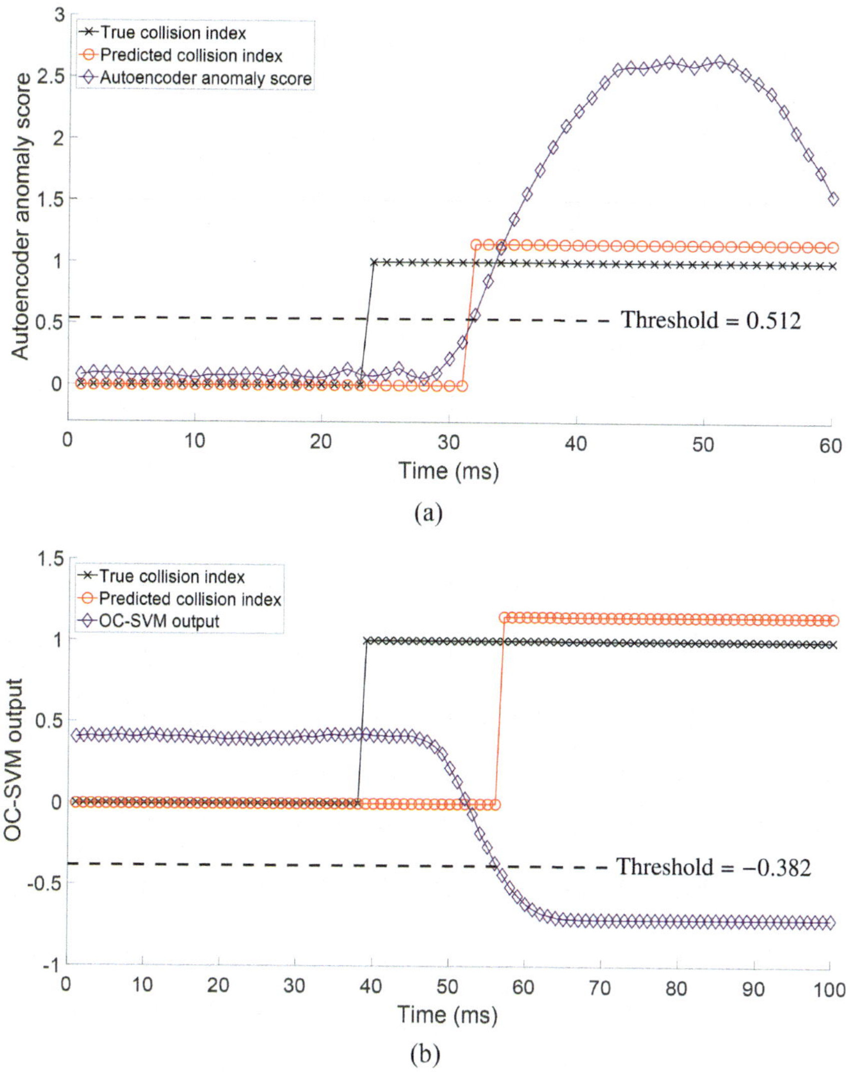

Fig. 2.5 Examples of detection output thresholding. **a** The non-negative scalar anomaly score resulting from the autoencoder-based reconstruction increases when a collision occurs. **b** The output scalar value of the trained OC-SVM decreases when a collision occurs

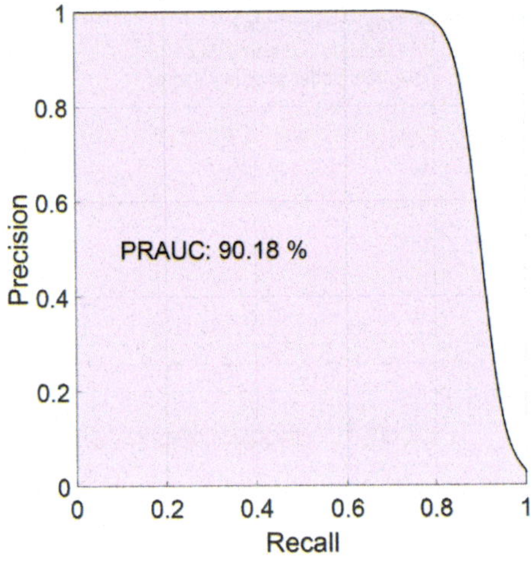

Fig. 2.6 The precision-recall curve with varying detection thresholds and its PRAUC of the autoencoder-based method

recall curve with varying detection thresholds and its PRAUC. We note that the optimal threshold that is derived by observing the collision-free motions (2.21) is equal to the threshold value that corresponds to the largest recall while precision is 1.

2.3.3.2 Detection Delay, Number of Detection Failures, and Number of False Alarms

Although PRAUC is useful for checking the global performance of a classifier with varying thresholds, in the case of collision detection with a fixed threshold, there is a limit to verifying the detection performance with only precision and recall. In collision detection problems, detecting collisions as fast as possible, i.e., with minimum detection delays, is more important than just correctly classifying the input samples. Figure 2.7 illustrates an example of a collision detected by two detection methods, where the predicted collision indices are denoted as predicted collision 1 and 2, respectively. Predicted collision 2 has a larger recall, but the collision is detected slower than predicted collision 1, showing that detection method 1 is more effective in detecting collisions although having a smaller recall. Therefore, for detecting collisions with a fixed threshold, we measure the average *detection delay* and count the *number of false alarms*.

The detection delay is defined by the sum of *propagation delay* and *inference delay*. As illustrated in Fig. 2.8, propagation delay is the time required for an impact caused by a collision to propagate through the robot inertia from the contact point to each joint, and inference delay is the time required for the collision detection algorithm to detect the collision from the robot joint signals. If the detection algorithm runs on a separate computer and not on the proprietary robot controller, *communication delay* between the computer and the controller should be added to the detection delay defined above. The detection delay we report in this book is only the sum of

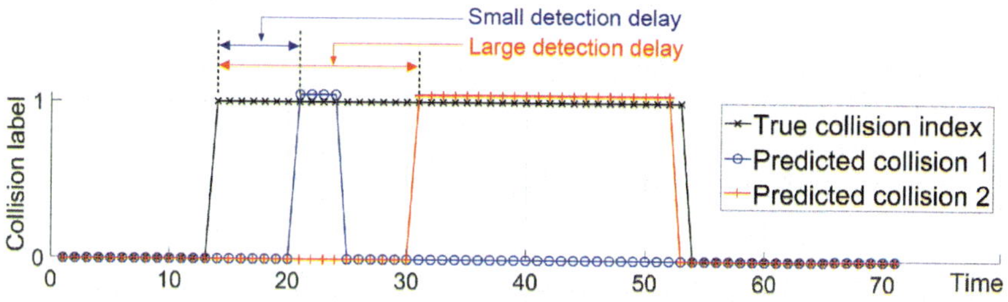

Fig. 2.7 An example of a collision detected by two detection methods

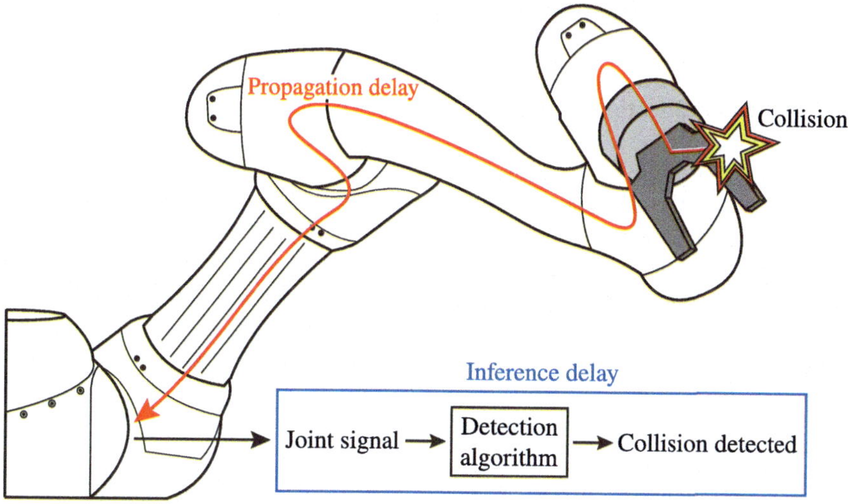

Fig. 2.8 The propagation delay and the inference delay

propagation delay and inference delay (communication delay is not considered). For situations in which the detection delay is longer than a certain predetermined maximum allowable delay t_{max} (in this book, we set $t_{max} = 250$ ms), we count them as *detection failures*. Here, we can also infer that the detection delay is lower bounded by the propagation delay and upper bounded by the maximum allowable delay t_{max}.

2.3.3.3 Computation Time (Real-Time Inference)

For a robot to operate in real-time, all aspects of the operating system including collision detection must run within the cycle time of the robot's control loop, e.g., 1 ms (1 kHz) or 250 µs (4 kHz). When a pure model-based method is selected for monitoring collisions, less care is required assuming the algorithm is properly implemented. For example, a basic momentum observer implemented in C++ on the proprietary controller of the collaborative robot manipulator *M0609* shows an average computation time of less than 1 μs. In the case of using a learning-based method that requires forward propagation of neural networks (the network training is done offline) or some other inference calculation [e.g., (2.36) for an OC-SVM], caution must be taken when designing the architecture of the network (including network input dimension) or other hyperparameters. If one designs a network too deep in order to improve performance (of course, this does not necessarily increase the performance), the forward propagation of the network alone may exceed the maximum allowable computation time. It is recommended to impose a computational limit on the network before tuning the hyperparameters. As long as the total computation does not exceed the maximum allowable computation, several detection methods may be used in parallel to increase the detection performance.

2.3.3.4 Generalization Capability

In the case of using learning techniques for detecting collisions, an important factor to be checked is the generalization capability of the trained network (or the learning algorithm). The trained network should not only detect well for the trained cases but also unseen situations. The generalization capability to be checked of a learning-based method may include generalization to e.g., various robot motions, unseen payload, joint angle range (configurations outside the trained range), different robot platforms, or aging of the robot and sensors (model errors and measurement noise).

2.4 Related Learning Algorithms

2.4.1 Support Vector Machine Regression

Given a training dataset $\{(x_1, y_1), \ldots, (x_N, y_N)\} \in \mathcal{X} \times \mathbb{R}$, where $\mathcal{X}$ is the input space and $\mathbb{R}$ is the output space, the goal of linear SVMR is to find a set of weights $w \in \mathcal{X}$ and a bias $b \in \mathbb{R}$ such that the linear function

$$f(x) = \langle w, x \rangle + b \tag{2.24}$$

is the best fit for the training data. One practical implementation is the following convex optimization:

$$\begin{aligned} &\underset{w,b}{\text{minimize}} && \frac{1}{2}\|w\|^2 \\ &\text{subject to} && \begin{cases} y_i - \langle w, x_i \rangle - b \leq \varepsilon \\ \langle w, x_i \rangle + b - y_i \leq \varepsilon \end{cases} \quad \forall i, \end{aligned} \tag{2.25}$$

for some user-prescribed threshold $\varepsilon \geq 0$. Since the constraints may not always be feasible, a soft-margin loss function is typically introduced using a set of slack variables $\xi_i, \xi_i^* \in \mathbb{R}$:

$$\begin{aligned} \underset{w,b,\xi,\xi^*}{\text{minimize}} \quad & \frac{1}{2}\|w\|^2 + C\sum_{i=1}^{N}(\xi_i + \xi_i^*) \\ \text{subject to} \quad & \begin{cases} y_i - \langle w, x_i\rangle - b \leq \varepsilon + \xi_i \\ \langle w, x_i\rangle + b - y_i \leq \varepsilon + \xi_i^* \\ \xi_i, \xi_i^* \geq 0 \end{cases} \quad \forall i, \end{aligned} \tag{2.26}$$

where the constants $C > 0$ and ε are user-defined parameters.

The corresponding dual version of (2.26) is

$$\begin{aligned} \underset{\alpha,\alpha^*\in\mathbb{R}^N}{\text{maximize}} \quad & \begin{cases} -\frac{1}{2}\sum\limits_{i,j=1}^{N}(\alpha_i - \alpha_i^*)(\alpha_j - \alpha_j^*)\langle x_i, x_j\rangle \\ -\varepsilon\sum\limits_{i=1}^{N}(\alpha_i + \alpha_i^*) + \sum\limits_{i=1}^{N} y_i(\alpha_i - \alpha_i^*) \end{cases} \\ \text{subject to} \quad & \begin{cases} \sum\limits_{i=1}^{N}(\alpha_i - \alpha_i^*) = 0 \\ \alpha_i, \alpha_i^* \in [0, C] \end{cases} \quad \forall i, \end{aligned} \tag{2.27}$$

where $\alpha_i, \alpha_i^* \in \mathbb{R}$ are the Lagrange multipliers. Given a sufficient amount of training data, a solution to the above quadratic programming problem can be obtained efficiently using, e.g., sequential minimal optimization [7, 8], with the required memory linear in the dataset size.

In terms of the solution to this optimization, (2.24) can be written as

$$f(x) = \sum_{i=1}^{N}(\alpha_i - \alpha_i^*)\langle x_i, x\rangle + b, \tag{2.28}$$

where w is now expressed as a linear combination of the training data x_i. From the Karush-Kuhn-Tucker (KKT) conditions, only a few of the $(\alpha_i - \alpha_i^*)$'s are non-zero, with the corresponding x_i's referred to as support vectors. This sparsity allows for good generalization to unseen test inputs as well as efficient computation. We refer the reader to [9] for further details.

The linear SVMR can be generalized to nonlinear regression problems by replacing the inner products $\langle x_i, x\rangle$ in (2.27) and (2.28) with the kernel function $k(x_i, x) = \langle \Phi(x_i), \Phi(x)\rangle$, where $\Phi : \mathcal{X} \to \mathcal{F}$ is a map from the input space $\mathcal{X}$ to some feature space $\mathcal{F}$. Any function $k(\cdot, \cdot)$ that satisfies Mercer's condition can be used without the knowledge of the map Φ. We refer to [10] for further details on the design of the kernel function.

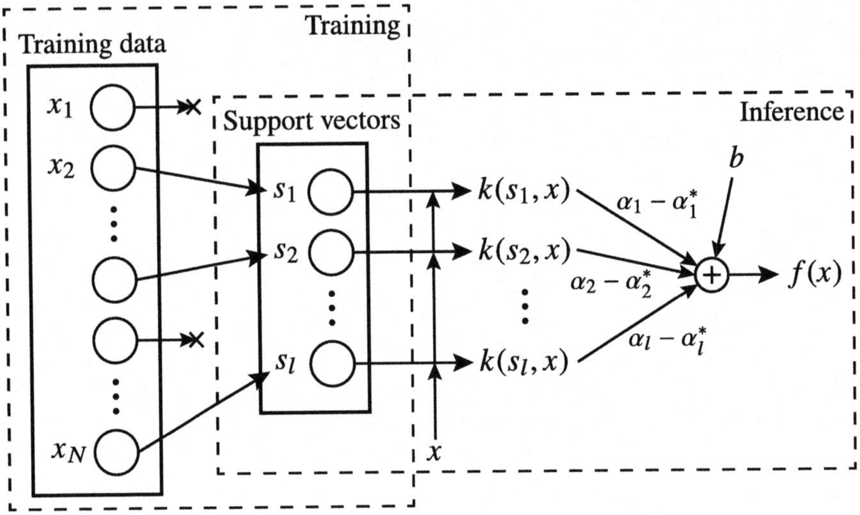

Fig. 2.9 A schematic diagram of support vector machine regression

A schematic diagram of SVMR is illustrated in Fig. 2.9. During the training phase, a few of the training data are chosen as support vectors, s_i, and the weights $\alpha_i - \alpha_i^*$ and bias b are determined. When a test input x is given, the output value is calculated by

$$f(x) = \sum_{i=1}^{l} (\alpha_i - \alpha_i^*) k(s_i, x) + b, \tag{2.29}$$

where l is the number of support vectors.

2.4.2 *One-Dimensional Convolutional Neural Network*

Convolutional neural network (CNN) is one of the most popular deep neural network architectures, commonly used for image recognition and classification [11]. Compared to simple fully-connected neural networks (FNNs), CNNs have three special concepts: local convolution, pooling layers, and filter-weight sharing. Due to these concepts, CNNs show better performance in extracting local features of strong 2-D structures, i.e., images, with fewer weight parameters than simple FNNs. Furthermore, 1-D CNNs have also been proposed for handling time-series data such as raw speech signals [12]; although recurrent neural networks are often used for time-series data, 1-D CNNs have also been applied to obtain remarkable results due to the strong 1-D structure of the data.

Figure 2.10 illustrates a one-dimensional convolution layer, where the 1-D convolution C is a map

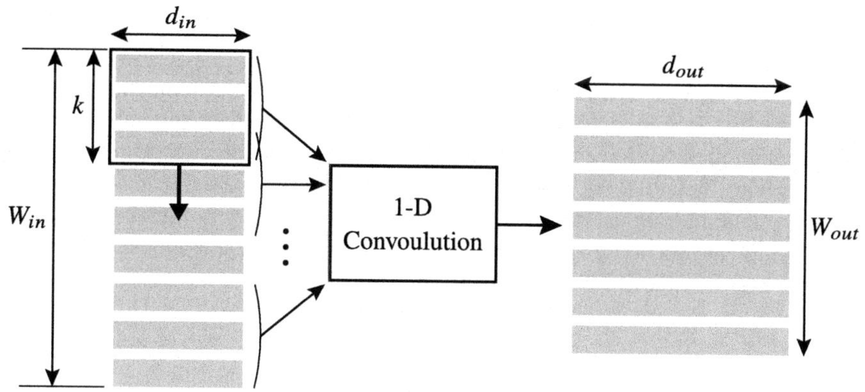

Fig. 2.10 A one-dimensional convolution layer

$$C : \mathbb{R}^{W_{in} \times d_{in}} \rightarrow \mathbb{R}^{W_{out} \times d_{out}}; \quad Z_{in} \mapsto Z_{out}, \tag{2.30}$$

where Z_{in} is the input matrix and Z_{out} is the output matrix of the layer. A *kernel* with kernel size k moves along the W-axis direction of the input matrix Z_{in} with a suitable *stride* d_s while applying a set of convolution operations with d_{out} convolution filters of size $k \times d_{in}$. The set of convolution operations maps k rows of Z_{in} to the corresponding row of Z_{out}. If a convolution layer utilizes the *same* padding, $2 \times p$ zero-filled extra rows are attached to the top and bottom of Z_{in} to preserve the window size W_{in}. On the other hand, if a convolution layer utilizes the *valid* padding, the window size W_{in} of the input is reduced to the window size W_{out} of the output. The output window size W_{out} is calculated as

$$W_{out} = \frac{W_{in} + 2 \times p - k}{d_s} + 1. \tag{2.31}$$

2.4.3 Unsupervised Anomaly Detection

Anomalies, also known as outliers or novelties, are observations that differ significantly from some concept of normality, which can be termed unusual, faulty, rare, or just simply strange. Various methods for detecting such anomalies using a set of observed data have been developed, in which the majority of the methods are unsupervised since labeled anomalous data are often not available. The main concept of unsupervised anomaly detection is learning the model of normality from a given dataset composed of normal observations so that anomalies can be detected through deviations from the trained model. Existing anomaly detection approaches can be organized into three main groups based on the used model: probabilistic models, one-class classification, and reconstruction models, in which all three groups formulate shallow and deep models. We refer the reader to [13, 14] for a seminal and comprehensive survey of anomaly detection methods.

In this book, we consider an OC-SVM, the most well-known one-class classification method, and an autoencoder, the most widely used deep anomaly detection model. Both methods have a long history and have been applied to various applications, proving their effectiveness; the flexible and expressive models can be applied to nonlinear and multimodal data and at the same time require very light computation. In the following sections, we briefly describe the principles and anomaly detection methodologies of the OC-SVM and the autoencoder.

2.4.4 One-Class Support Vector Machine

We consider a training dataset $\{x_1, \ldots, x_N\} \in \mathcal{X}$, where $\mathcal{X}$ is some input space, and a kernel function $k(x_i, x_j) = \langle \Phi(x_i), \Phi(x_j) \rangle$, where $\Phi : \mathcal{X} \to \mathcal{F}$ is a map from the input space $\mathcal{X}$ to some feature space $\mathcal{F}$. The goal of an OC-SVM [15] is to find the hyperplane in the feature space that separates the mapped training dataset and the origin with maximum margin. The hyperplane can be expressed in the form $\langle w, \Phi(x) \rangle - \rho = 0$, where $w \in \mathcal{F}$ and $\rho \in \mathbb{R}$. Then, the decision function

$$f(x) = \langle w, \Phi(x) \rangle - \rho \tag{2.32}$$

is positive for a small region containing most of the training data and negative elsewhere. One practical implementation is as the following quadratic programming:

$$\begin{aligned} \underset{w \in \mathcal{F}, \xi \in \mathbb{R}^N, \rho \in \mathbb{R}}{\text{minimize}} \quad & \frac{1}{2} \|w\|^2 + \frac{1}{\nu N} \sum_{i=1}^{N} \xi_i - \rho \\ \text{subject to} \quad & \begin{cases} \langle w, \Phi(x_i) \rangle \geq \rho - \xi_i \\ \xi_i \geq 0 \end{cases} \quad \forall i, \end{aligned} \tag{2.33}$$

where the constant $\nu \in (0, 1]$ is a user-defined parameter.

The corresponding dual problem of (2.33) is

$$\begin{aligned} \underset{\alpha \in \mathbb{R}^N}{\text{minimize}} \quad & \frac{1}{2} \sum_{i,j=1}^{N} \alpha_i \alpha_j k(x_i, x_j) \\ \text{subject to} \quad & \begin{cases} 0 \leq \alpha_i \leq \frac{1}{\nu N} \\ \sum_{i=1}^{N} \alpha_i = 1, \end{cases} \quad \forall i \end{aligned} \tag{2.34}$$

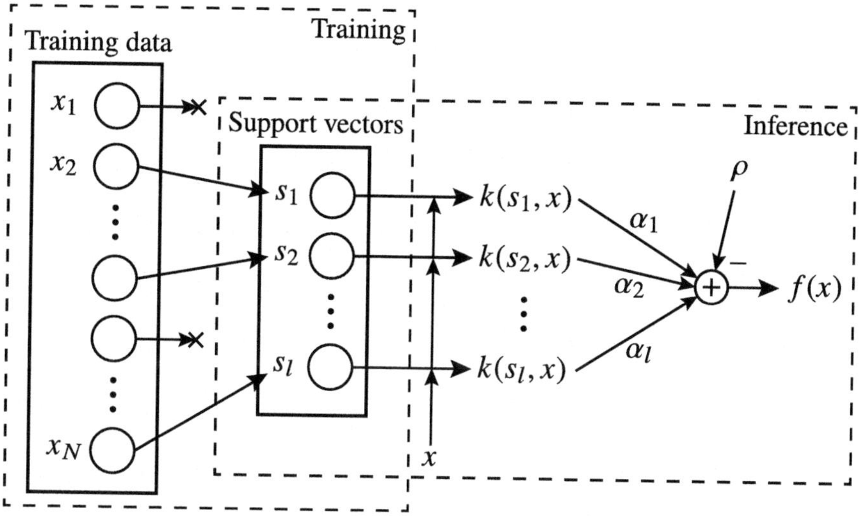

Fig. 2.11 A schematic diagram of a one-class support vector machine

where $\alpha \in \mathbb{R}^N$ is the Lagrange multiplier. Given a sufficient amount of training data, a solution to the above dual problem can be obtained efficiently using, e.g., sequential minimal optimization [7, 8], with the required memory linear in the dataset size.

In terms of the solution to this optimization, (2.32) can be written as

$$f(x) = \sum_{i=1}^{N} \alpha_i k(x_i, x) - \rho, \tag{2.35}$$

where w is now expressed as a linear combination of the mapped training data $\Phi(x_i)$. From the KKT conditions, only a few of the α_i's are non-zero, with the corresponding x_i's referred to as the support vectors. This sparsity allows for good generalization to unseen test inputs as well as efficient computation.

A schematic diagram of an OC-SVM is illustrated in Fig. 2.11. During the training phase, a few of the training data are chosen as support vectors, s_i, and the weights α_i and bias ρ are determined. When an unseen test input x is given, the output value is calculated by

$$f(x) = \sum_{i=1}^{l} \alpha_i k(s_i, x) - \rho, \tag{2.36}$$

where l is the number of support vectors; $f(x) \geq 0$ indicates that x is a normal data, and $f(x) < 0$ indicates that x is an anomalous data.

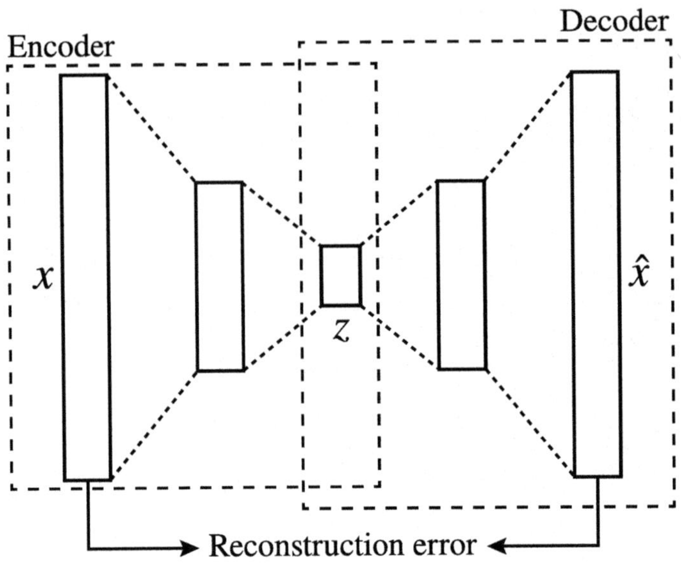

Fig. 2.12 A schematic diagram of an autoencoder

2.4.5 *Autoencoder-Based Anomaly Detection*

2.4.5.1 Autoencoder Network Architecture and Training

An autoencoder is a multilayer neural network used for reducing the dimensionality of data [16]. The network is composed of the encoder part, which transforms the high-dimensional input data into a low-dimensional representation code, and the decoder part, which reconstructs the input data from the dimension-reduced code. As illustrated in Fig. 2.12, the input and output layers have the same number of nodes, while the central hidden layer has a smaller number of nodes. The network can be trained by minimizing the reconstruction error defined by

$$e(x) = \|x - \hat{x}\|^2 \in \mathbb{R}, \tag{2.37}$$

where x is the original input data and $\hat{x}$ is the reconstructed output. If the dimensionality reduction is successful, the discrepancy between the original input and the reconstructed output is considerably small.

2.4.5.2 Anomaly Detection Using Autoencoders

An autoencoder can be used for anomaly detection by learning the normal data [17]. During the training phase, the autoencoder is optimized to reconstruct the normal training inputs, and therefore the trained autoencoder cannot effectively reconstruct anomalous inputs that are not similar to the training data, resulting in high reconstruction errors. By using the reconstruction error as the anomaly score, anomalies can be effectively detected.

References

1. Lynch, K.M., Park, F.C.: Modern Robotics. Cambridge University Press, Cambridge (2017)
2. Spong, M.W.: Modeling and control of elastic joint robots. ASME J. Dyn. Syst. Measur. Control **109**(4), 310–318 (1987)
3. De Luca, A., Book, W.J.: Robots with flexible elements. In: Springer Handbook of Robotics, pp. 243–282. Springer, Berlin (2016)
4. Haddadin, S., De Luca, A., Albu-Schäffer, A.: Robot collisions: a survey on detection, isolation, and identification. IEEE Trans. Robot. **33**(6), 1292–1312 (2017)
5. Park, K.M., Kim, J., Park, J., Park, F.C.: Learning-based real-time detection of robot collisions without joint torque sensors. IEEE Rob. Autom. Lett. **6**(1), 103–110 (2020)
6. Park, K.M., Park, Y., Yoon, S., Park, F.C.: Collision detection for robot manipulators using unsupervised anomaly detection algorithms. IEEE/ASME Trans. Mechatron. (2021)
7. Platt, J.: Sequential minimal optimization: a fast algorithm for training support vector machines. Technical Report MSR-TR-98-14 (1998)
8. Fan, R.-E., Chen, P.-H., Lin, C.-J.: Working set selection using second order information for training support vector machines. J. Mach. Learn. Res. **6**(Dec.), 1889–1918 (2005)
9. Smola, A.J., Schölkopf, B.: A tutorial on support vector regression. Stat. Comput. **14**(3), 199–222 (2004)
10. Genton, M.G.: Classes of kernels for machine learning: a statistics perspective. J. Mach. Learn. Res. **2**(Dec.), 299–312 (2001)
11. LeCun, Y., Bottou, L., Bengio, Y., Haffner, P., et al.: Gradient-based learning applied to document recognition. Proc. IEEE **86**(11), 2278–2324 (1998)
12. Palaz, D., Collobert, R., Doss, M.M.: Estimating phoneme class conditional probabilities from raw speech signal using convolutional neural networks. In: Proceedings of Interspeech, pp. 1766–1770 (2013)
13. Goldstein, M., Uchida, S.: A comparative evaluation of unsupervised anomaly detection algorithms for multivariate data. PloS One **11**(4), e0152173 (2016)
14. Ruff, L., Kauffmann, J.R., Vandermeulen, R.A., Montavon, G., Samek, W., Kloft, M., Dietterich, T.G., Müller, K.-R.: A unifying review of deep and shallow anomaly detection. Proc. IEEE (2021)
15. Schölkopf, B., Platt, J.C., Shawe-Taylor, J., Smola, A.J., Williamson, R.C.: Estimating the support of a high-dimensional distribution. Neural Comput. **13**(7), 1443–1471 (2001)
16. Rumelhart, D.E., Hinton, G.E., Williams, R.J.: Learning internal representations by error propagation. Technical Report 8506, California University, San Diego, La Jolla, Institute for Cognitive Science (1985)
17. Japkowicz, N., Myers, C., Gluck, M., et al.: A novelty detection approach to classification. IJCAI **1**, 518–523 (1995). Citeseer

Chapter 3
Model-Free and Model-Based Methods

3.1 Introduction

Existing collision detection methods can be loosely divided into two categories depending on whether the dynamics model of the robot is used (model-based methods) or not (model-free methods). In this chapter, we first introduce some model-free detection methods that typically rely on additional external sensors. Then, we describe basic model-based methods, where we follow the organization and derivation methodologies of [1], together with pseudocode descriptions of the algorithms for real system implementation. Also, methods to deal with the frictional torque in the joints (which is critical when detecting collisions) and some additional alternative model-based methods are reviewed.

3.2 Model-Free Methods

Model-free methods typically rely on additional exteroceptive sensors attached to or placed around the manipulator. The first type of sensor that comes to mind for monitoring collisions is the tactile surface sensor. Surface sensors show a very fast response time to external force and have the further advantage of not only detecting collisions or contacts but also measuring the exact contact point. Also, some sensors even have features for measuring the exact applied force, providing flexibility to the reaction strategies. Due to these desirable features, various types of tactile surface sensors have been developed for collision detection and physical human-robot interaction [2–8]. Three-axis accelerometers are also used for detecting collisions, where the vibration that occurs during an impact is analyzed to detect and estimate the location and direction of the collisions [9–12]. These methods are effective in detecting hard collisions and have the further advantage of identifying the collided material, which leads to the capability of impact classification of accidental collisions and intentional

K. M. Park and F. C. Park, *Collision Detection for Robot Manipulators: Methods and Algorithms*, Springer Tracts in Advanced Robotics 155,
https://doi.org/10.1007/978-3-031-30195-7_3

contacts. Recently, detection systems based on acoustic information collected from microphones attached to the robot have been proposed, where these methods show similar features to those of the three-axis accelerometer-based approaches [13, 14]. Not common in the collision detection phase, but RGB-D cameras placed around the robot can also be used for monitoring physical collisions [15, 16]. By using multiple cameras placed around the robot, it is possible to detect potential collisions given the complete trajectory that the robot is executing.

Despite the desirable features of the above methods, there are limitations that cannot be ignored including the significant increase in system complexity. It is difficult to cover the entire surface of the robot with the tactile sensors, and repeated hard collisions can easily damage the sensors. Methods using accelerometers and microphones have limitations in detecting soft collisions, e.g., pushing and pulling motions. In the case of using cameras, the view from multiple cameras must be calibrated, and the computation is heavy, requiring a GPU. From both a cost and performance perspective, model-based methods that rely on a robot dynamics model together with joint signals have been more successful than model-free methods.

3.3 Basic Model-Based Methods

For the model-based methods, we first describe four basic methods based on rigid joint robot dynamics: (1) energy observer-based estimation of the external power, and estimation of the external joint torque using (2) direct estimation, (3) a velocity observer, and (4) a momentum observer. We refer the reader to [1] for a more in-depth discussion of the above four methods. Then, methods to deal with the frictional torque in the joints are reviewed, e.g., friction modeling, using a force-torque sensor, and using joint torque sensors, followed by some additional alternative model-based methods, e.g., model-compensating methods and IMU sensor-based.

3.3.1 Energy Observer-Based Estimation of External Power

The external power resulting from a collision can be monitored by using an energy observer. Given the positive gain for the observer, $k_O \in \mathbb{R}$, the energy observer dynamics is defined as

$$\dot{\hat{E}} = \dot{q}^T(\tau_m - \tau_F) + r_e, \tag{3.1}$$

$$\dot{r}_e = k_O(\dot{E} - \dot{\hat{E}}). \tag{3.2}$$

The observer output $r_e(t) \in \mathbb{R}$ of the energy observer is described by the equation

$$r_e = k_O\left(E(t) - \int_0^t (\dot{q}^T(\tau_m - \tau_F) + r_e)ds - E(0)\right) \tag{3.3}$$

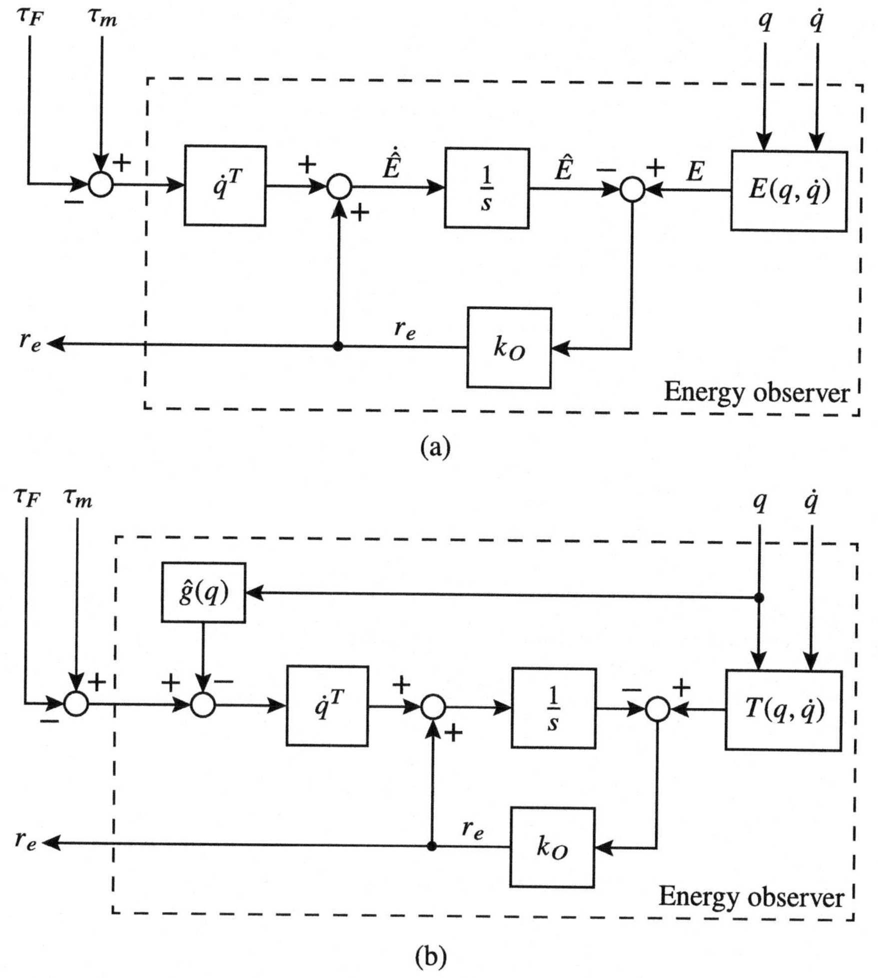

Fig. 3.1 **a** The block diagram of an energy observer, and **b** its equivalent form

with $E(q, \dot{q}) = T(q, \dot{q}) + \hat{U}(q) = \frac{1}{2}\dot{q}^T \hat{M}(q)\dot{q} + \hat{U}(q)$, where $\hat{M}$ denotes the nominal model parameter and $\hat{U}$ is the estimate of the potential energy [17]. Figure 3.1a illustrates the block diagram of an energy observer. Under ideal condition where $\hat{E} = E$, the dynamic relation between $r_e(t)$ and the external power $P_{ext} = \dot{q}^T \tau_{ext}$ is

$$\begin{aligned} \dot{r}_e &= k_O(\dot{E} - \dot{\hat{E}}) \\ &= k_O(\dot{q}^T(\tau_m + \tau_{ext} - \tau_F) - (\dot{q}^T(\tau_m - \tau_F) + r_e)) \\ &= k_O(\dot{q}^T \tau_{ext} - r_e) \\ &= k_O(P_{ext} - r_e), \end{aligned} \tag{3.4}$$

so that $r_e(t)$ acts as an estimate for P_{ext} (more precisely, the first order low-pass filtered P_{ext}) where

$$r_e(t) = P_{ext}(1 - e^{-k_O t}). \tag{3.5}$$

Since $\dot{U}(q) = \dot{q}^T g(q)$, it can be seen that

$$\hat{U}(t) - \hat{U}(0) = \int_0^t \dot{\hat{U}} ds = \int_0^t \dot{q}^T \hat{g}(q) ds, \tag{3.6}$$

where $\hat{g}$ denote the nominal model parameter, and with (3.6), we can derive the equivalent form of (3.3) as

$$r_e = k_O \left(T(t) - \int_0^t (\dot{q}^T (\tau_m - \tau_F - \hat{g}(q)) + r_e) ds - T(0) \right). \tag{3.7}$$

By using (3.7), computation of $\hat{U}(q)$ can be replaced by $\hat{g}(q)$. Figure 3.1b illustrates the block diagram of the equivalent energy observer.

The estimated external power is compared with a user-defined fixed threshold $\epsilon_{cd} \in \mathbb{R}$, so that the collision decision function $cd : r_e(t) \rightarrow \{True, False\}$ is defined as

$$cd(r_e(t)) = \begin{cases} True & \text{if } |r_e(t)| > \epsilon_{cd} \\ False & \text{if } |r_e(t)| \leq \epsilon_{cd} \end{cases}. \tag{3.8}$$

The binary output indicates whether or not a collision has occurred.

Assuming that the initial velocity and the initial potential energy are set to zero ($\dot{q}(0) = 0, \hat{U}(0) = 0$), with Δt denoting the cycle time of the robot's control loop, the energy observer-based collision detection algorithm can be written as Algorithm 1 or Algorithm 2, respectively based on (3.3) and (3.7).

The energy observer-based detection method has the advantages of being simple to implement and requiring only one detection threshold. However, the major limitation of using energy observers is that there are situations where collisions cannot be detected (even if the external force is sufficiently high). First, if the robot is initially at rest before a collision occurs, since $\dot{q} = 0$, τ_{ext} does not affect P_{ext} and therefore r_e is also not affected. Also from (2.12), we can see that if the collision wrench $\mathcal{F}_c$ is orthogonal to the collision twist $\mathcal{V}_c$, P_{ext} remains zero.

Algorithm 1 Collision detection with an energy observer

$r_e \leftarrow 0$
$\hat{E} \leftarrow 0$
$cd \leftarrow False$
while $t > 0$ **do**
  $\hat{M}(q) \leftarrow \text{ComputeMassMatrix}(q)$
  $\hat{U}(q) \leftarrow \text{ComputePotentialEnergy}(q)$
  $\hat{E} \leftarrow \hat{E} + (\dot{q}^T(\tau_m - \tau_F) + r_e) \times \Delta t$
  $E \leftarrow \frac{1}{2}\dot{q}^T \hat{M}(q)\dot{q} + \hat{U}(q)$
  $r_e \leftarrow k_O(E - \hat{E})$
  if $|r_e| > \epsilon_{cd}$ **then**
    $cd \leftarrow True$
  end if
  if $cd = True$ **then**
    break
  end if
end while
Reaction to collision

Algorithm 2 Collision detection with an energy observer (equivalent form)

$r_e \leftarrow 0$
$\hat{T} \leftarrow 0$
$cd \leftarrow False$
while $t > 0$ **do**
  $\hat{M}(q) \leftarrow \text{ComputeMassMatrix}(q)$
  $\hat{g}(q) \leftarrow \text{ComputeGravityVector}(q)$
  $\hat{T} \leftarrow \hat{T} + (\dot{q}^T(\tau_m - \tau_F - \hat{g}(q)) + r_e) \times \Delta t$
  $T \leftarrow \frac{1}{2}\dot{q}^T \hat{M}(q)\dot{q}$
  $r_e \leftarrow k_O(T - \hat{T})$
  if $|r_e| > \epsilon_{cd}$ **then**
    $cd \leftarrow True$
  end if
  if $cd = True$ **then**
    break
  end if
end while
Reaction to collision

3.3.2 Direct Estimation of External Joint Torque

Estimating the external joint torque is useful not only for detecting collisions but also for identifying the contact point or the collided link (the isolation phase), and for estimating the impact force (the identification phase). The most straightforward and theoretically ideal method of estimating the external joint torque is to directly calculate τ_{ext} from (2.2) as

$$\hat{\tau}_{ext} = \hat{M}(q)\ddot{q} + \hat{C}(q, \dot{q})\dot{q} + \hat{g}(q) + \tau_F - \tau_m, \tag{3.9}$$

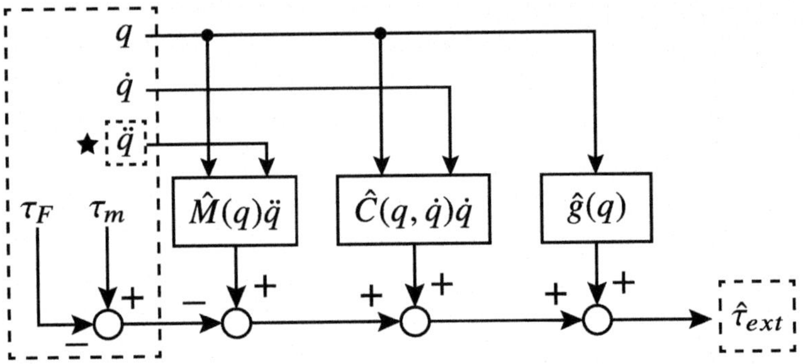

Fig. 3.2 The block diagram of the direct estimation-based method

where $\hat{\tau}_{ext}$ is the estimate of the external joint torque and $\hat{M}(q)$, $\hat{C}(q, \dot{q})$, and $\hat{g}(q)$ are the nominal model parameters.

Provided an accurate estimate or measurement of the joint acceleration $\ddot{q}$, the output of the direct estimation-based method (Fig. 3.2) immediately responds to a collision without delay while the velocity and momentum observer-based methods exhibit inevitable delays since the observer outputs exponentially approach τ_{ext} (3.14), (3.21). However, accurate measurement of $\ddot{q}$ is often not available, and if $\ddot{q}$ is estimated by differentiating $\dot{q}$ or double differentiating q, the estimated $\ddot{q}$ may include non-negligible noise, requiring heavy filtering which leads to increments of detection failure rate and detection delay.

3.3.3 Velocity Observer-Based Estimation of External Joint Torque

For the velocity observer, we use the notation

$$n(q, \dot{q}) := C(q, \dot{q})\dot{q} + g(q) \tag{3.10}$$

for compactness. Given the positive-definite diagonal gain matrix for the observer, $K_O = \text{diag}\{k_{O,i}\} \in \mathbb{R}^{n \times n}$, the velocity observer dynamics is defined as

$$\hat{\ddot{q}} = \hat{M}^{-1}(q)(\tau_m - \tau_F - \hat{n}(q, \dot{q}) + r_v), \tag{3.11}$$

$$\dot{r}_v = K_O(\ddot{q} - \hat{\ddot{q}}). \tag{3.12}$$

The observer output $r_v(t) \in \mathbb{R}^n$ of the velocity observer is described by the equation

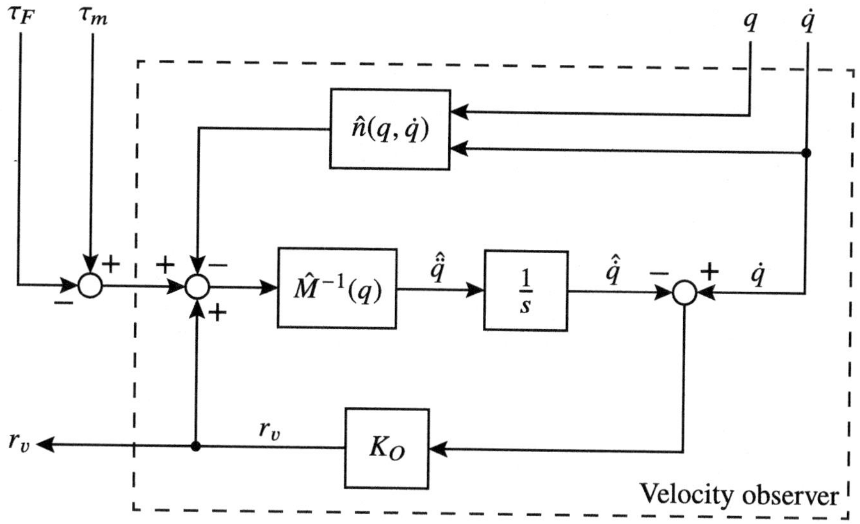

Fig. 3.3 The block diagram of a velocity observer

$$r_v = K_O \left(\dot{q}(t) - \int_0^t \hat{M}^{-1}(q) \left(\tau_m - \tau_F - \hat{n}(q, \dot{q}) + r_v \right) ds - \dot{q}(0) \right) \quad (3.13)$$

with $\hat{n} = \hat{C}(q, \dot{q})\dot{q} + \hat{g}(q)$, where $\hat{M}$, $\hat{C}$, and $\hat{g}$ denote the nominal model parameters [18]. We can see from (3.13) that when computing r_v, the computation of $\hat{M}^{-1}(q)$ is required. Figure 3.3 illustrates the block diagram of a velocity observer. Under ideal conditions where $\hat{M} = M$ and $\hat{n} = n$, the dynamic relation between $r_v(t)$ and the external joint torque τ_{ext} is

$$\begin{aligned} \dot{r}_v &= K_O(\ddot{q} - \hat{\ddot{q}}) \\ &= K_O \left(M^{-1}(q) (\tau_m + \tau_{ext} - \tau_F - n) - \hat{M}^{-1}(q) \left(\tau_m - \tau_F - \hat{n} + r_v \right) \right) \\ &= K_O M^{-1}(q)(\tau_{ext} - r_v), \end{aligned} \quad (3.14)$$

so that $r_v(t)$ acts as an estimate for τ_{ext} where

$$r_v(t) = (I - e^{-K_O M^{-1}(q)t})\tau_{ext}. \quad (3.15)$$

Unlike the energy observer-based method (3.4) or the momentum observer-based method (3.21), the sensitivity of the observer output r_v to the value to be estimated τ_{ext} is not constant; the estimation sensitivity depends on the robot's configuration.

Algorithm 3 Collision detection with a velocity observer

$r_v \leftarrow (0, \ldots, 0) \in \mathbb{R}^n$
$\hat{\dot{q}} \leftarrow (0, \ldots, 0) \in \mathbb{R}^n$
$cd \leftarrow False$
while $t > 0$ **do**
  $\hat{M}(q) \leftarrow \text{ComputeMassMatrix}(q)$
  $\hat{M}^{-1}(q) \leftarrow \text{MatrixInverse}(\hat{M}(q))$
  $\hat{C}(q, \dot{q}) \leftarrow \text{ComputeCoriolisMatrix}(q, \dot{q})$
  $\hat{g}(q) \leftarrow \text{ComputeGravityVector}(q)$
  $\hat{\dot{q}} \leftarrow \hat{\dot{q}} + \hat{M}^{-1}(q)(\tau_m - \tau_F - \hat{C}(q, \dot{q})\dot{q} - \hat{g}(q) + r_v) \times \Delta t$
  $r_v \leftarrow K_O(\dot{q} - \hat{\dot{q}})$
  for $j \leftarrow 1$ to n **do**
    if $|r_{v,j}| > \epsilon_{cd,j}$ **then**
      $cd \leftarrow True$
      break
    end if
  end for
  if $cd = True$ **then**
    break
  end if
end while
Reaction to collision

The estimated external joint torque is compared component-wisely with a user-defined fixed threshold $\epsilon_{cd} = (\epsilon_{cd,1}, \ldots, \epsilon_{cd,n}) \in \mathbb{R}^n$, so that the collision decision function $cd : r_v(t) \rightarrow \{True, False\}$ is defined as

$$cd(r_v(t)) = \begin{cases} True & \text{if } |r_v(t)| > \epsilon_{cd} \\ False & \text{if } |r_v(t)| \leq \epsilon_{cd} \end{cases}. \tag{3.16}$$

The binary output indicates whether or not a collision has occurred.

Assuming that the robot is initially at rest, i.e., $\dot{q}(0) = 0$, with $r_{v,j}$ and $\epsilon_{cd,j}$ denoting the jth component of r_v and ϵ_{cd}, respectively, and Δt denoting the cycle time of the robot's control loop, the velocity observer-based collision detection algorithm can be written as Algorithm 3.

3.3.4 *Momentum Observer-Based Estimation of External Joint Torque*

For the momentum observer, we use the notation

$$\beta(q, \dot{q}) := g(q) - C^T(q, \dot{q})\dot{q} \tag{3.17}$$

for compactness. Given the positive-definite diagonal gain matrix for the observer, $K_O = \text{diag}\{k_{O,i}\} \in \mathbb{R}^{n \times n}$, the momentum observer dynamics is defined as

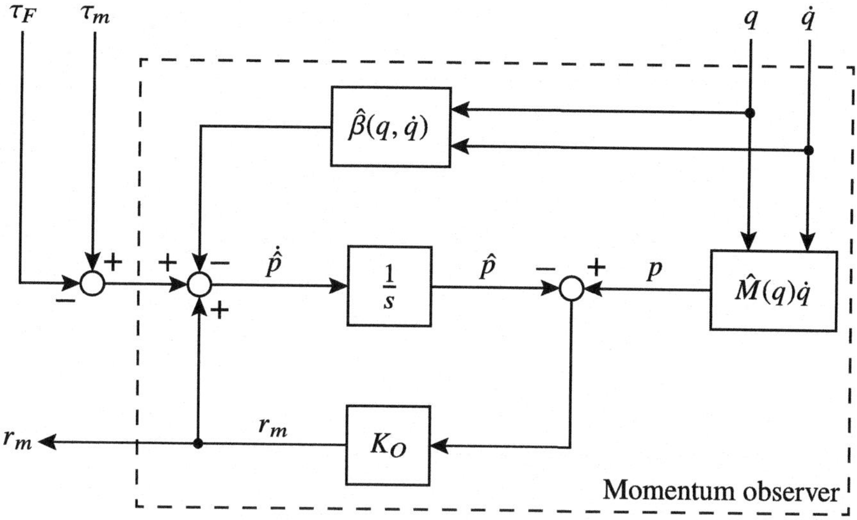

Fig. 3.4 The block diagram of a momentum observer

$$\dot{\hat{p}} = \tau_m - \tau_F - \hat{\beta}(q, \dot{q}) + r_m, \tag{3.18}$$

$$\dot{r}_m = K_O(\dot{p} - \dot{\hat{p}}). \tag{3.19}$$

The observer output $r_m(t) \in \mathbb{R}^n$ of the momentum observer is described by the equation

$$r_m = K_O \left(p(t) - \int_0^t (\tau_m - \tau_F - \hat{\beta} + r_m) ds - p(0) \right) \tag{3.20}$$

with $p = \hat{M}(q)\dot{q}$ and $\hat{\beta} = \hat{g}(q) - \hat{C}^T(q, \dot{q})\dot{q}$, where $\hat{M}$, $\hat{C}$, and $\hat{g}$ denote the nominal model parameters [17, 19]. Figure 3.4 illustrates the block diagram of a momentum observer. Under ideal conditions where $\hat{M} = M$ and $\hat{\beta} = \beta$, the dynamic relation between $r_m(t)$ and the external joint torque τ_{ext} is

$$\begin{aligned} \dot{r}_m &= K_O(\dot{p} - \dot{\hat{p}}) \\ &= K_O(\tau_m + \tau_{ext} - \tau_F - \beta(q, \dot{q}) - (\tau_m - \tau_F - \hat{\beta}(q, \dot{q}) + r_m)) \\ &= K_O(\tau_{ext} - r_m), \end{aligned} \tag{3.21}$$

so that $r_m(t)$ acts as an estimate for τ_{ext} where

$$r_m(t) = (I - e^{-K_O t})\tau_{ext}. \tag{3.22}$$

The estimated external joint torque is compared component-wisely with a user-defined fixed threshold $\epsilon_{cd} = (\epsilon_{cd,1}, \ldots, \epsilon_{cd,n}) \in \mathbb{R}^n$, so that the collision decision function $cd : r_m(t) \rightarrow \{True, False\}$ is defined as

$$cd(r_m(t)) = \begin{cases} True & \text{if } |r_m(t)| > \epsilon_{cd} \\ False & \text{if } |r_m(t)| \leq \epsilon_{cd} \end{cases}. \tag{3.23}$$

The binary output indicates whether or not a collision has occurred.

Assuming that the robot starts at rest, i.e., $\dot{q}(0) = 0$, with $r_{m,j}$ and $\epsilon_{cd,j}$ denoting the jth component of r_m and ϵ_{cd}, respectively, and Δt denoting the cycle time of the robot's control loop, the momentum observer-based collision detection algorithm can be written as Algorithm 4.

Algorithm 4 Collision detection with a momentum observer

$r_m \leftarrow (0, \ldots, 0) \in \mathbb{R}^n$
$\hat{p} \leftarrow (0, \ldots, 0) \in \mathbb{R}^n$
$cd \leftarrow False$
while $t > 0$ **do**
  $\hat{M}(q) \leftarrow$ ComputeMassMatrix(q)
  $\hat{C}(q, \dot{q}) \leftarrow$ ComputeCoriolisMatrix($q, \dot{q}$)
  $\hat{C}^T(q, \dot{q}) \leftarrow$ MatrixTranspose($\hat{C}(q, \dot{q})$)
  $\hat{g}(q) \leftarrow$ ComputeGravityVector(q)
  $\hat{p} \leftarrow \hat{p} + (\tau_m - \tau_F - \hat{g}(q) + \hat{C}^T(q, \dot{q})\dot{q} + r_m) \times \Delta t$
  $p \leftarrow \hat{M}(q)\dot{q}$
  $r_m \leftarrow K_O(p - \hat{p})$
  for $j \leftarrow 1$ to n **do**
    if $|r_{m,j}| > \epsilon_{cd,j}$ **then**
      $cd \leftarrow True$
      break
    end if
  end for
  if $cd = True$ **then**
    break
  end if
end while
Reaction to collision

While several different external joint torque estimation methods have been described, the general consensus is that the momentum observer-based method is so far the most accurate, reliable, and cost-effective; the estimation sensitivity is independent of the robot's configuration and has the further advantage of not requiring joint acceleration estimates, the inverse of the mass matrix, and any other additional exteroceptive sensors. Due to these desirable features, the momentum observer-based method is known to be the most widely used method to reliably monitor collisions, and there are several works that try to improve the performance of the momentum observer [20–22]. Also, the momentum observer output is used as the network input for many learning-based detection methods [23–26].

3.4 Dealing with Frictional Torque in the Joints

One of the challenges with using model-based detection methods is to deal with the dissipative frictional torque τ_F in the joints. For example, assume that a momentum observer is used for collision monitoring without considering the frictional torque. Then, the output $r \in \mathbb{R}^n$ of the observer is

$$r = K_O \left(p(t) - \int_0^t (\tau_m - \hat{\beta} + r)ds - p(0) \right), \tag{3.24}$$

where the dynamics of the observer is defined as (3.18) without the τ_F term and (3.19). The relation between $r(t)$ and the external joint torque τ_{ext} becomes

$$\dot{r} = K_O(\tau_{ext} - \tau_F - r), \tag{3.25}$$

$r(t)$ now acting as an estimate for $\tau_{ext} - \tau_F$ rather than for τ_{ext}. If the detection threshold is set as (2.21) in this situation, the threshold would be too large since the observer output does not exhibit values near zero even when there is no collision, resulting in multiple detection failures (especially for hard collisions).

3.4.1 *Friction Modeling and Identification*

The first approach one can consider for addressing this issue is to accurately model and identify the friction parameters. The friction is usually modeled as a function of joint velocity $\dot{q}$, e.g., as a (third-order) polynomial function of $\dot{q}$ [27], as the sum of an affine function and a sigmoidal function [28], or as the sum of multiple hyperbolic tangential functions [29]. There also exist friction models that consider both q and $\dot{q}$, such as in [30]. Equations (3.26), (3.27), (3.28), and (3.29) show the simplified version of the key parts of the friction models proposed in [27–29], and [30] respectively:

$$[27] : \gamma_1 \dot{q}^3 + \gamma_2 \dot{q}^2 + \gamma_3 \dot{q}, \tag{3.26}$$

$$[28] : (\gamma_1 \dot{q} + \gamma_2) + \gamma_3 \left(1 + e^{-\gamma_4(\dot{q}+\gamma_5)}\right)^{-1}, \tag{3.27}$$

$$[29] : \gamma_1 \left(\tanh(\gamma_2 \dot{q}) - \tanh(\gamma_3 \dot{q})\right) + \gamma_4 \tanh(\gamma_5 \dot{q}) + \gamma_6 \dot{q}, \tag{3.28}$$

$$[30] : \gamma_1 + \gamma_2 \dot{q} + \gamma_3 \exp(-\gamma_4 \dot{q}^2) + \gamma_5 \sin q + \gamma_6 \cos q + \gamma_7 \sin 2q + \gamma_8 \cos 2q, \tag{3.29}$$

where $\gamma_i \in \mathbb{R}$ are the friction parameters. After collecting the fitting data from sample trajectories, the friction parameters are identified offline by the least squares technique.

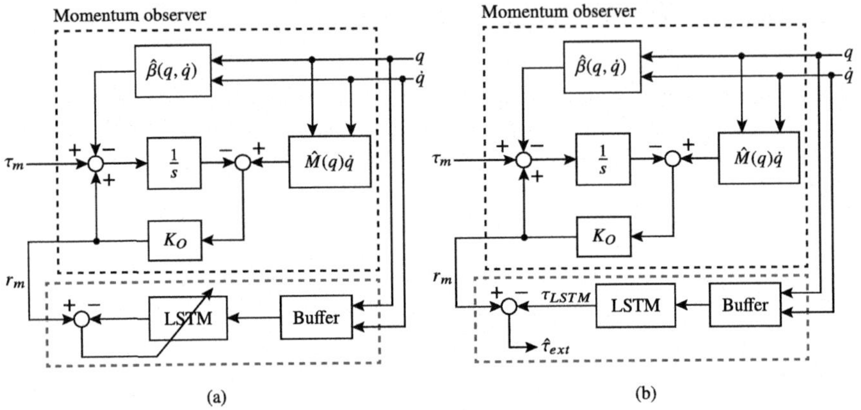

Fig. 3.5 The block diagram of the framework proposed in [32]. **a** Disturbance torque learning during collision-free motions. **b** Friction compensation with the trained LSTM during collision monitoring

3.4.2 Learning Friction

Recently, some works proposed methods that use neural networks to automatically learn (model and identify) the frictional torque. As we can see from (3.25), a disturbance observer, e.g., the momentum observer, can be used for estimating the frictional torque (and also the disturbance torque resulting from model uncertainties) during collision-free motions. The observer output is then learned using a neural network, where the input is a combination of q and $\dot{q}$, and the network output is the frictional torque. In [31], an FNN is used to learn the frictional torque estimated by a nonlinear disturbance observer, and the proposed algorithm is validated for a one-joint motion. In [32], a Long Short-Term Memory (LSTM) is trained to learn the disturbance torque estimated by a momentum observer during collision-free motions (Fig. 3.5a). By subtracting the output τ_{LSTM} of the trained LSTM from the momentum observer output, the friction-compensated external joint torque is estimated (Fig. 3.5b). Implementation and validation are performed for two-joint vertical motions with various modeling errors (e.g., errors in the mass and the center of mass position).

Instead of learning the friction model separately and subtracting it from the observer output, some methods propose networks that compensate for the friction internally. In [24, 25] (which are described in detail in Chaps. 4 and 5), the friction-contaminated momentum observer output (3.24) is used as the input for the learning algorithms. As shown in Fig. 3.6, the scalar output of the learning algorithm (SVMR) stays near zero when no external force is applied and increases only when a collision is detected, even without separate compensation for the non-negligible frictional torque. In [26], in addition to the observer output, the joint velocity $\dot{q}$ is given as input to the network. By doing so, the network automatically learns the nonlinear friction model internally and compensates for the external joint torque (as described in Sect. 3.4.1, the friction is usually modeled as a function of $\dot{q}$).

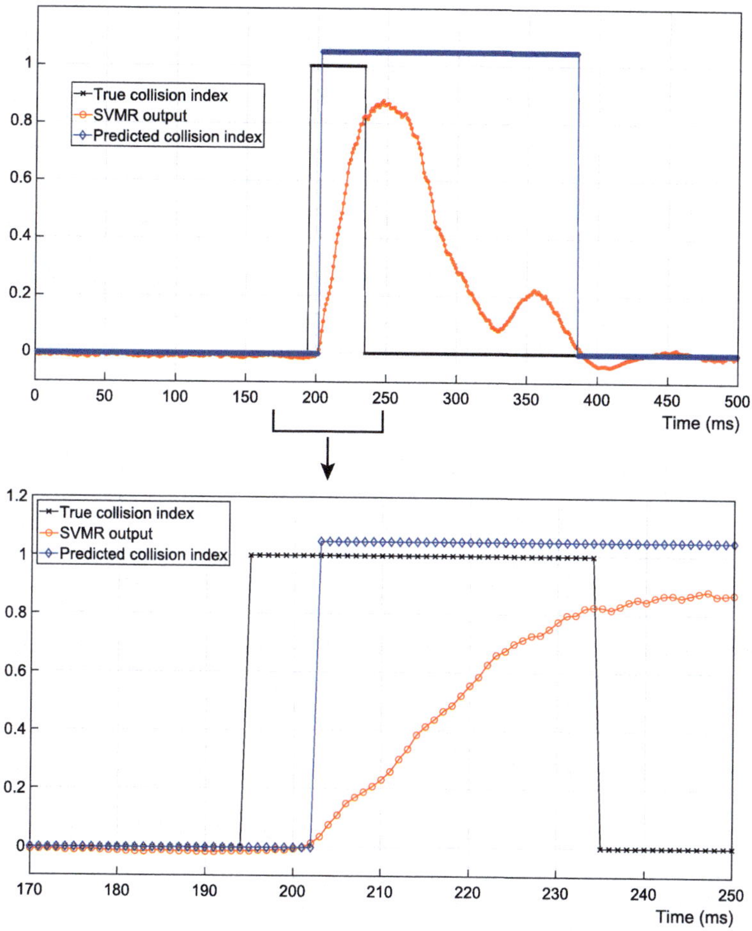

Fig. 3.6 The scalar output of the trained SVMR stays near zero when no external force is applied

3.4.3 Using Additional Sensors

However, since friction is nonlinear and complex (especially during velocity reversal) and it changes with the payload of the robot [33], finding or learning a perfectly matching friction model can be challenging. Some alternative methods can be applied to avoid the modeling of friction when additional sensors are available. In [34], a collision monitoring method requiring a force-torque (F/T) sensor installed at the robot's base (bedplate) as abstractly illustrated in Fig. 3.7 has been proposed. A dynamic model that estimates the reaction F/T at the base caused by the manipulator's dynamics (dynamic force) is modeled and identified. Then, the external F/T is calculated by subtracting the estimated dynamic force from the reaction F/T mea-

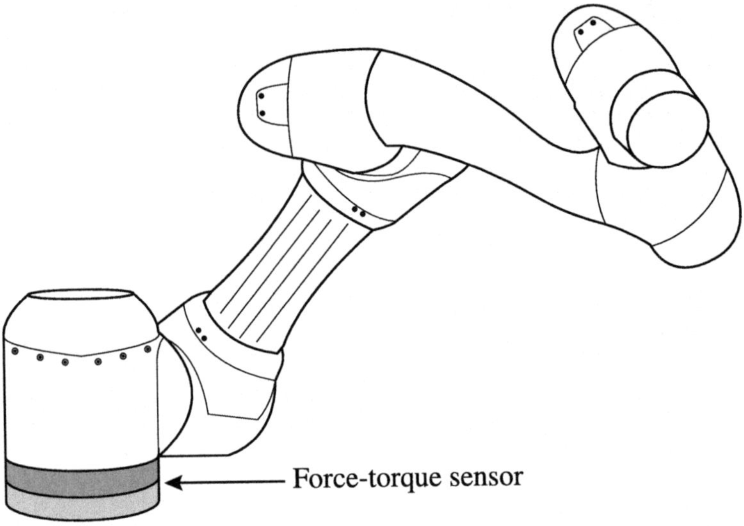

Fig. 3.7 A force-torque sensor installed at the base of a 6-DOF robot manipulator

sured by the sensor. Collision is monitored by comparing the calculated external F/T with a prescribed six-dimensional (three for force and three for torque) threshold.

Another approach is to attach joint torque sensors to the joints and use the link-side momentum observer [1]. As can be seen from the link dynamics (2.13) and the motor dynamics (2.14) of the robot, the two dynamics are coupled by the elastic joint torque τ_J. If τ_J is accurately measured by the joint torque sensors, the two dynamics can be decoupled, where the motor-side frictional torque τ_F only appears in the motor dynamics. The link-side momentum observer takes the advantage of this decoupling feature, where the observer dynamics is defined as

$$\dot{\hat{p}}_q = \tau_J - \hat{\beta}(q, \dot{q}) + r_J, \tag{3.30}$$

$$\dot{r}_J = K_O(\dot{p}_q - \dot{\hat{p}}_q). \tag{3.31}$$

The observer output $r_J(t) \in \mathbb{R}^n$ is described by the equation

$$r_J = K_O \left(p_q(t) - \int_0^t (\tau_J - \hat{\beta} + r_J) ds - p_q(0) \right) \tag{3.32}$$

with $p_q = \hat{M}(q)\dot{q}$ as illustrated in Fig. 3.8. The differences between (3.20) and (3.32) are that the active motor torque τ_m is replaced by the measured joint torque τ_J and the frictional torque term τ_F is removed. Under ideal conditions, the dynamic relation between $r_J(t)$ and the external joint torque τ_{ext} is

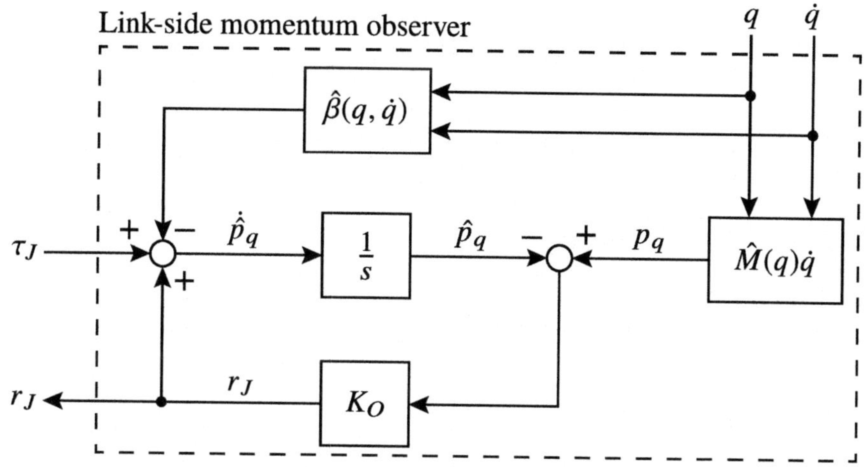

Fig. 3.8 The block diagram of a link-side momentum observer

$$\begin{aligned}\dot{r}_J &= K_O(\dot{p}_q - \dot{\hat{p}}_q) \\ &= K_O(\tau_J + \tau_{ext} - \beta(q, \dot{q}) - (\tau_J - \hat{\beta}(q, \dot{q}) + r_J)) \\ &= K_O(\tau_{ext} - r_J),\end{aligned} \tag{3.33}$$

so that $r_J(t)$ acts as an estimate for τ_{ext} even without knowing τ_F. We note that the model-free approaches that use exteroceptive sensors are not mentioned here since those methods do not use the dynamics model in the first place and therefore are free from the frictional torque.

3.5 Collision Detection Under Cyclic Motions

Under the special case of a robot performing cyclic (repetitive) motions, the *cycle normalization* technique [23] can be applied to improve the detection performance. Assume that a robot is repeating the same motion every period T before a collision occurs, which is the normal operating mode. During the normal operating mode, the monitoring signal $x(t) \in \mathbb{R}^k$ (which can include either model-free or model-based signals) of one cycle (usually the first) is stored as the reference cycle signal:

$$X_{ref} = [x(0), x(\Delta t), \ldots, x(T)]. \tag{3.34}$$

Each time a new cycle starts, the reference cycle signal is subtracted from the present monitoring signal as

$$\overline{X}_t = X_t - X_{ref}, \tag{3.35}$$

where X_t is the present cycle signal and $\overline{X}_t$ is the cycle-normalized signal. The synchronization of the two signals can be done by, for example, simply comparing the time index [23], or using a technique entitled Dynamic Time Warping which finds the optimal matching of compared signals [35].

The normalized signal can be directly used for collision detection when provided with proper detection thresholds or can be used as the network input (for both training and inference) for learning-based methods. When used for collision learning, the trained network may effectively generalize to different robot platforms or various cyclic robot motions [23] (the generalization capability of learning-based methods is a notable issue). In addition, although there are no relevant research results yet (to the best of our knowledge), we believe the cycle normalization technique can also be used for friction compensation in cyclic motions. The frictional torque is generally known to be dependent on the robot state q and $\dot{q}$, payload, temperature, and time [1]. When a robot is performing a repetitive motion, (ideally) the same robot state will be repeated every cycle and therefore each cycle will exhibit a similar frictional torque profile. By normalizing the monitoring signal as (3.35), the friction is expected to be compensated. Further studies are required.

3.6 Alternative Model-Based Methods

There are also other existing model-based methods that improve detection performance in many ways. As introduced in Sect. 3.3.4, there are several works based on the momentum observer [20–22]. In [20], the authors derive new observers based on the generalized momentum dynamics that (theoretically) achieve finite-time convergence while the basic momentum observer estimates a first-order filtered external joint torque. The same issue is also addressed in [21], presenting a nonlinear extended state momentum observer which eliminates the initial peak that occurs during the estimation and improves detection accuracy. The method proposed in [22] utilizes two channels, one sensitive to hard collisions and the other performing better for soft collisions, improving the bandwidth of the momentum observer.

When using model-based methods, not only friction but also the parameters of the dynamics model have to be accurately identified. Even after proper identification, however, there may still exist unmodeled effects or uncertainties in the model, e.g., backlash or elasticity. Therefore, some methods have been proposed to compensate for these model uncertainties. In [36], the residual between the applied and the estimated torque is high-pass filtered and compared with a set of state-dependent dynamic thresholds, where the filter coefficients and the dynamic thresholds are estimated online by adaptive estimation. Implementation and validation of the algorithm are performed for hard collisions occurring in two-joint motions (which are restricted to the vertical plane) of a flexible joint manipulator. The algorithm presented in [37] also addresses the issue of parametric uncertainties in the robot manipulator, based on the frequency characteristic difference between the collisions and modeling errors.

As mentioned in Sect. 3.3.2, immediate response to a collision can be realized by the direct estimation-based method if an accurate estimate or measurement of the joint acceleration $\ddot{q}$ is available (an accurate frictional torque model is also needed). The authors of [38] implement the direct estimation-based method at a practically relevant level for the first time. Their method requires at least one IMU sensor to be attached per robot link. By fusing the joint position and the IMU measurements, the joint acceleration (and also the velocity) is accurately estimated.

3.7 Multiple Arms and End-Effector Types

All the methods described so far are developed for a single open-chain robot manipulator. Also, we did not consider the end-effector, which varies in type according to its application, e.g., grippers for grabbing objects, magnets for manipulating ferrous materials, and vacuum heads for moving non-porous-surfaced parts. Figure 3.9 shows a situation where two (or more) robot manipulators are installed with different types of end-effectors attached. Implementing collision detection for this case is straightforward. Since an impact from a collision propagates through the robot inertia from its contact point to the robot's base (as illustrated in Fig. 2.8), an external force acting on one robot does not affect the other robot. In fact, finding the collided link with the estimated external joint torque (for single contact case) is based on this feature, where the collided link is identified by the last joint among the joints exhibiting non-zero external joint torques [1]. Therefore, all we have to do is to implement two collision monitoring systems, each responsible for one robot. Also due to this feature, as long as the end-effector does not absorb all the impact applied to it, the impact will affect the joint signal (e.g., motor current, joint torque) and the implemented detection system will report the collision. We note that in such an environment, studies to prevent collision between the robots in the first place (self-collision avoidance) are being actively conducted [39–42].

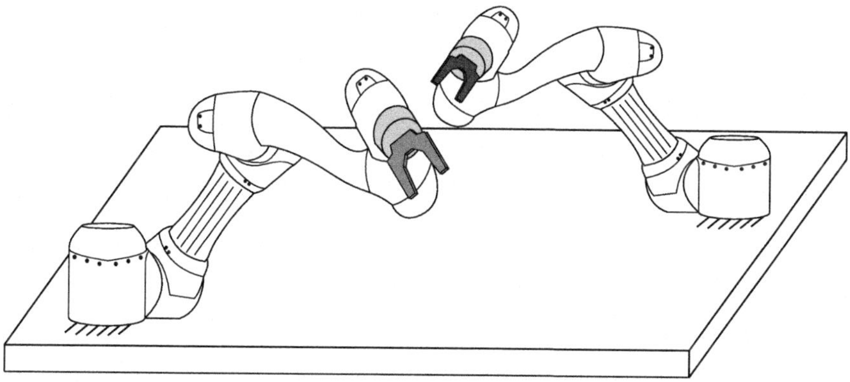

Fig. 3.9 Two robot manipulators with different types of end-effectors attached

3.8 Summary

In this chapter, we have first introduced some model-free detection methods that rely on additional external sensors—tactile surface sensors, three-axis accelerometers, microphones, and cameras. These methods show desirable features not only for detection but also for isolation, identification, and classification. Then, we described the four basic model-based detection methods—energy observer-based, direct estimation-based, velocity observer-based, and momentum observer-based. Among the four methods, the general consensus is that the momentum observer-based method is so far the most accurate, reliable, and cost-effective. Next, several approaches for dealing with the frictional torque in the joints have been introduced—friction modeling, learning friction, and using additional sensors. Collision detection under the special case of the robot performing cyclic motions has also been discussed. Finally, after introducing some alternative model-based methods (improving the performance of the momentum observer, dynamics model compensating methods, and joint acceleration estimation), extensions to multiple arms and to other end-effector types have been discussed.

References

1. Haddadin, S., De Luca, A., Albu-Schäffer, A.: Robot collisions: a survey on detection, isolation, and identification. IEEE Trans. Rob. **33**(6), 1292–1312 (2017)
2. Duchaine, V., Lauzier, N., Baril, M., Lacasse, M.-A., Gosselin, C.: A flexible robot skin for safe physical human robot interaction. In: 2009 IEEE International Conference on Robotics and Automation, pp. 3676–3681. IEEE (2009)
3. Strohmayr, M.W., Wörn, H., Hirzinger, G.: The DLR artificial skin step I: uniting sensitivity and collision tolerance. In: 2013 IEEE International Conference on Robotics and Automation, pp. 1012–1018. IEEE (2013)
4. Cirillo, A., Ficuciello, F., Natale, C., Pirozzi, S., Villani, L.: A conformable force/tactile skin for physical human-robot interaction. IEEE Rob. Autom. Lett. **1**(1), 41–48 (2015)
5. Hughes, D., Lammie, J., Correll, N.: A robotic skin for collision avoidance and affective touch recognition. IEEE Rob. Autom. Lett. **3**(3), 1386–1393 (2018)
6. Pang, G., Yang, G., Heng, W., Ye, Z., Huang, X., Yang, H.-Y., Pang, Z.: CoboSkin: soft robot skin with variable stiffness for safer human-robot collaboration. IEEE Trans. Ind. Electron. **68**(4), 3303–3314 (2020)
7. Ye, Z., Pang, G., Xu, K., Hou, Z., Lv, H., Shen, Y., Yang, G.: Soft robot skin with conformal adaptability for on-body tactile perception of collaborative robots. IEEE Rob. Autom. Lett. **7**(2), 5127–5134 (2022)
8. Ji, S., Shin, J., Yoon, J., Lim, K.-H., Sim, G.-D., Lee, Y.-S., Kim, D.H., Cho, H., Park, J.: Three-dimensional skin-type triboelectric nanogenerator for detection of two-axis robotic-arm collision. Nano Energy **97**, 107225 (2022)
9. Wisanuvej, P., Liu, J., Chen, C.-M., Yang, G.-Z.: Blind collision detection and obstacle characterisation using a compliant robotic arm. In: 2014 IEEE International Conference on Robotics and Automation (ICRA), pp. 2249–2254. IEEE (2014)
10. Min, F., Wang, G., Liu, N.: Collision detection and identification on robot manipulators based on vibration analysis. Sensors **19**(5), 1080 (2019)
11. Sandykbayeva, D., Kappassov, Z., Orazbayev, B.: VibroTouch: active tactile sensor for contact detection and force sensing via vibrations. Sensors **22**(17), 6456 (2022)

12. Valle, C.M.C.O., Kurdas, A., Fortunić, E.P., Abdolshah, S., Haddadin, S.: Real-time IMU-based learning: a classification of contact materials. In: 2022 IEEE/RSJ International Conference on Intelligent Robots and Systems (IROS), pp. 1965–1971. IEEE (2022)
13. Fan, X., Lee, D., Chen, Y., Prepscius, C., Isler, V., Jackel, L., Sebastian Seung, H., Lee, D.: Acoustic collision detection and localization for robot manipulators. In: 2020 IEEE/RSJ International Conference on Intelligent Robots and Systems (IROS), pp. 9529–9536. IEEE (2020)
14. Dimiccoli, M., Patni, S., Hoffmann, M., Moreno-Noguer, F.: Recognizing object surface material from impact sounds for robot manipulation. In: 2022 IEEE/RSJ International Conference on Intelligent Robots and Systems (IROS), pp. 9280–9287. IEEE (2022)
15. Maric, B., Jurican, F., Orsag, M., Kovacic, Z.: Vision based collision detection for a safe collaborative industrial manipulator. In: 2021 IEEE International Conference on Intelligence and Safety for Robotics (ISR), pp. 334–337. IEEE (2021)
16. Makris, S., Aivaliotis, P.: AI-based vision system for collision detection in HRC applications. Procedia CIRP **106**, 156–161 (2022)
17. De Luca, A., Albu-Schaffer, A., Haddadin, S., Hirzinger, G.: Collision detection and safe reaction with the DLR-III lightweight manipulator arm. In: 2006 IEEE/RSJ International Conference on Intelligent Robots and Systems, pp. 1623–1630. IEEE (2006)
18. Haddadin, S.: Towards Safe Robots: Approaching Asimov's 1st Law, vol. 90. Springer, Berlin (2013)
19. De Luca, A., Mattone, R.: Sensorless robot collision detection and hybrid force/motion control. In: Proceedings of the 2005 IEEE International Conference on Robotics and Automation, pp. 999–1004. IEEE (2005)
20. Garofalo, G., Mansfeld, N., Jankowski, J., Ott, C.: Sliding mode momentum observers for estimation of external torques and joint acceleration. In: 2019 International Conference on Robotics and Automation (ICRA), pp. 6117–6123. IEEE (2019)
21. Li, Y., Li, Y., Zhu, M., Xu, Z., Mu, D.: A nonlinear momentum observer for sensorless robot collision detection under model uncertainties. Mechatronics **78**, 102603 (2021)
22. Huang, S., Gao, M., Liu, L., Chen, J., Zhang, J.: Collision detection for cobots: a back-input compensation approach. IEEE/ASME Trans. Mechatron. (2022)
23. Heo, Y.J., Kim, D., Lee, W., Kim, H., Park, J., Chung, W.K.: Collision detection for industrial collaborative robots: a deep learning approach. IEEE Rob. Autom. Lett. **4**(2), 740–746 (2019)
24. Park, K.M., Kim, J., Park, J., Park, F.C.: Learning-based real-time detection of robot collisions without joint torque sensors. IEEE Rob. Autom. Lett. **6**(1), 103–110 (2020)
25. Park, K.M., Park, Y., Yoon, S., Park, F.C.: Collision detection for robot manipulators using unsupervised anomaly detection algorithms. IEEE/ASME Trans. Mechatron. (2021)
26. Kim, D., Lim, D., Park, J.: Transferable collision detection learning for collaborative manipulator using versatile modularized neural network. IEEE Trans. Rob. (2021)
27. Lee, S.-D., Song, J.-B.: Sensorless collision detection based on friction model for a robot manipulator. Int. J. Precis. Eng. Manuf. **17**(1), 11–17 (2016)
28. Gaz, C., Magrini, E., De Luca, A.: A model-based residual approach for human-robot collaboration during manual polishing operations. Mechatronics **55**, 234–247 (2018)
29. Zhang, C., Mu, C., Wang, Y., Li, J., Liu, Z.: Collision detection for six-DOF serial robots force/position hybrid control based on continuous friction model. Meas. Control 00202940221091575 (2022)
30. Xiao, J., Zhang, Q., Hong, Y., Wang, G., Zeng, F.: Collision detection algorithm for collaborative robots considering joint friction. Int. J. Adv. Rob. Syst. **15**(4), 1729881418788992 (2018)
31. Xu, T., Fan, J., Fang, Q., Zhu, Y., Zhao, J.: A new robot collision detection method: a modified nonlinear disturbance observer based-on neural networks. J. Intell. Fuzzy Syst. **38**(1), 175–186 (2020)
32. Lim, D., Kim, D., Park, J.: Momentum observer-based collision detection using LSTM for model uncertainty learning. In: 2021 IEEE International Conference on Robotics and Automation (ICRA), pp. 4516–4522. IEEE (2021)
33. Swevers, J., Verdonck, W., De Schutter, J.: Dynamic model identification for industrial robots. IEEE Control Syst. Mag. **27**(5), 58–71 (2007)

34. Li, W., Han, Y., Wu, J., Xiong, Z.: Collision detection of robots based on a force/torque sensor at the bedplate. IEEE/ASME Trans. Mechatron. **25**(5), 2565–2573 (2020)
35. Gordić, Z., Jovanović, K.: Collision detection on industrial robots in repetitive tasks using modified dynamic time warping. Robotica **38**(10), 1717–1736 (2020)
36. Makarov, M., Caldas, A., Grossard, M., Rodriguez-Ayerbe, P., Dumur, D.: Adaptive filtering for robust proprioceptive robot impact detection under model uncertainties. IEEE/ASME Trans. Mechatron. **19**(6), 1917–1928 (2014)
37. Birjandi, S.A.B., Haddadin, S.: Model-adaptive high-speed collision detection for serial-chain robot manipulators. IEEE Rob. Autom. Lett. **5**(4), 6544–6551 (2020)
38. Birjandi, S.A.B., Kühn, J., Haddadin, S.: Observer-extended direct method for collision monitoring in robot manipulators using proprioception and IMU sensing. IEEE Rob. Autom. Lett. **5**(2), 954–961 (2020)
39. Fang, C., Rocchi, A., Hoffman, E.M., Tsagarakis, N.G., Caldwell, D.G.: Efficient self-collision avoidance based on focus of interest for humanoid robots. In: 2015 IEEE-RAS 15th International Conference on Humanoid Robots (Humanoids), pp. 1060–1066. IEEE (2015)
40. Fernandez, N.B.F., Salehian, S.S.M., Billard, A.: Multi-arm self-collision avoidance: a sparse solution for a big data problem. In: Proceedings of the Third Machine Learning in Planning and Control of Robot Motion (MLPC) Workshop, number CONF (2018)
41. Marinho, M.M., Adorno, B.V., Harada, K., Mitsuishi, M.: Active constraints using vector field inequalities for surgical robots. In: 2018 IEEE International Conference on Robotics and Automation (ICRA), pp. 5364–5371. IEEE (2018)
42. Quiroz-Omaña, J.J., Adorno, B.V.: Whole-body control with (self) collision avoidance using vector field inequalities. IEEE Rob. Autom. Lett. **4**(4), 4048–4053 (2019)

Chapter 4
Learning Robot Collisions

4.1 Introduction

Recently, learning-based methods for collision detection have gained popularity as a means of sidestepping some of the implementation difficulties of pure model-based methods and compensating for uncertain dynamic effects, e.g., unmodeled or uncertain parameters in the robot dynamics model and measurement noise. Instead of friction modeling and identification, a learning network is trained with a set of robot collision training data (which will be described in detail below). Provided a sufficient amount of diverse, good-quality data has been collected, the trained network can then compensate for uncertainties in the dynamics model parameters as well as unmodeled effects like friction and measurement noise. Furthermore, for some methods, the manual tuning process of multiple detection thresholds can be eliminated. Based on the learning algorithms used, we can divide existing learning-based detection methods into two types: one that uses supervised learning algorithms and the other that uses unsupervised learning algorithms.

In this chapter, we first introduce some existing supervised learning-based collision detection methods. Then, to show step-by-step how a learning-based detection algorithm is implemented (which may seem unfamiliar yet), we describe in detail the two methods we have developed—one based on SVMR, and the other using a 1-D CNN [1]. The methods are easy to implement and show desirable features in terms of detection performance, computation (network complexity), and generalization capability. After presenting the robot collision dataset we have collected, we describe in detail for each method the procedure for input feature design, the training methodology of the learning algorithms, and the collision detection (inference) methodology using the trained learning algorithms. The detection performance and the generalization capability of the methods are also verified for the test dataset.

K. M. Park and F. C. Park, *Collision Detection for Robot Manipulators: Methods and Algorithms*, Springer Tracts in Advanced Robotics 155,
https://doi.org/10.1007/978-3-031-30195-7_4

4.2 Supervised Learning-Based Approaches

Supervised learning-based methods typically require the true labels of the training data, which include binary true collision index and torque values (frictional or external). These methods can be categorized into two groups based on the required collision scenarios of the training data: one that requires data for both collision and collision-free motions of the robot, and the other that requires only the data for collision-free motions.

4.2.1 Training With Only Collision-Free Data

In fact, the methods which only use collision-free motion data for training have already been introduced in Sect. 3.4.2. The frictional or the disturbance torque estimated by a disturbance observer is learned during collision-free motions and does not require the data for collisions [2, 3]. Among the two methods, we will describe [3], which is validated for more general circumstances (two-joint robot motions with various modeling errors), in a little more detail.

The monitoring signal which is selected as the true label of the LSTM network's training data is the output of the momentum observer that does not consider the frictional torque (3.24). As described before, the observer output $r(t)$ estimates $\tau_{ext} - \tau_F$ (3.25). However, since the network is trained for only collision-free motions, the external joint torque is identically zero, i.e., $\tau_{ext}(t) \equiv 0$. Then, the observer acts as an estimator for the disturbance torque

$$\tau_{dist} = -\tau_F, \tag{4.1}$$

where we assume that the model uncertainty torque is included in τ_F for simplicity of notation.

The authors utilize a many-to-one LSTM network (with 1 hidden layer and 20 hidden units) as illustrated in Fig. 4.1 which takes sequential data as input and outputs a one-dimensional value, which in this case is the estimated disturbance torque. The history of the states q (which represents the robot configuration) and $\dot{q}$ are given as the network input

$$X(t) = [x(t - 99\Delta t), \ldots, x(t - \Delta t), x(t)], \tag{4.2}$$

where

$$x(t) = [q(t), \dot{q}(t), \dot{q}(t - \Delta t)]. \tag{4.3}$$

After the network is trained, the network output τ_{LSTM} acts as the estimated disturbance torque which is then subtracted from the observer output for friction (and model

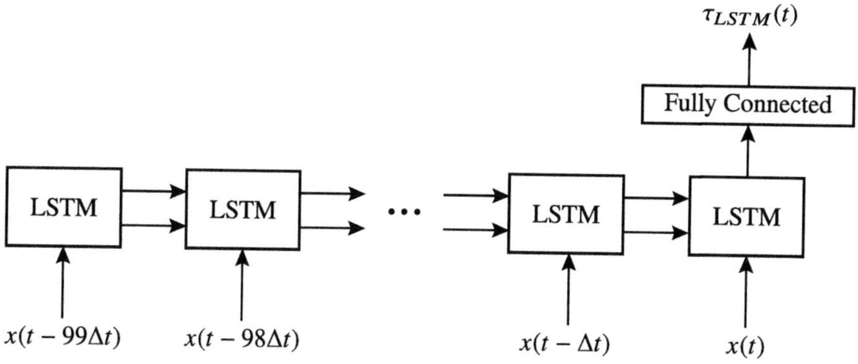

Fig. 4.1 The LSTM network architecture proposed in [3]

uncertainty) compensation. With extensive experiments using a 2-DOF vertical-structured robot, the authors verify that the trained network successfully estimates the disturbance torque.

4.2.2 Training with both Collision and Collision-Free Data

Most of the existing supervised learning-based detection methods use both collision and collision-free data for training. Based on the output of the utilized learning algorithm, these methods can be categorized into one of three schemes: (1) estimating the external joint torque, (2) end-to-end networks that output a binary predicted collision index, and (3) learning algorithms that output a scalar collision prediction score. The first approach estimates the external joint torque τ_{ext} without (or with only a part of) the dynamics model by using a neural network. The n-dimensional network output is then thresholded component-wisely as (2.19). The second type of method directly outputs the predicted collision index $\{True, False\}$, eliminating the thresholding process. The network automatically learns the threshold internally, which is desirable when manual sensitivity adjustment is unnecessary. The 1-D CNN-based method we have developed belongs to this group. The third approach, where the SVMR-based method belongs, uses learning algorithms that output a scalar collision prediction score which is then thresholded as (2.20). The distinction between the second and the third approaches may seem to be ambiguous since the end-to-end networks also eventually require thresholding at the final step, where the detection threshold is usually set to very close to 1, e.g., 0.99. In this book, only the methods that (1) exhibit a (relatively) gradually increasing output when a collision occurs and (2) the false alarms can be effectively suppressed with a threshold value not close to 1, e.g., 0.1, are categorized as the third group.

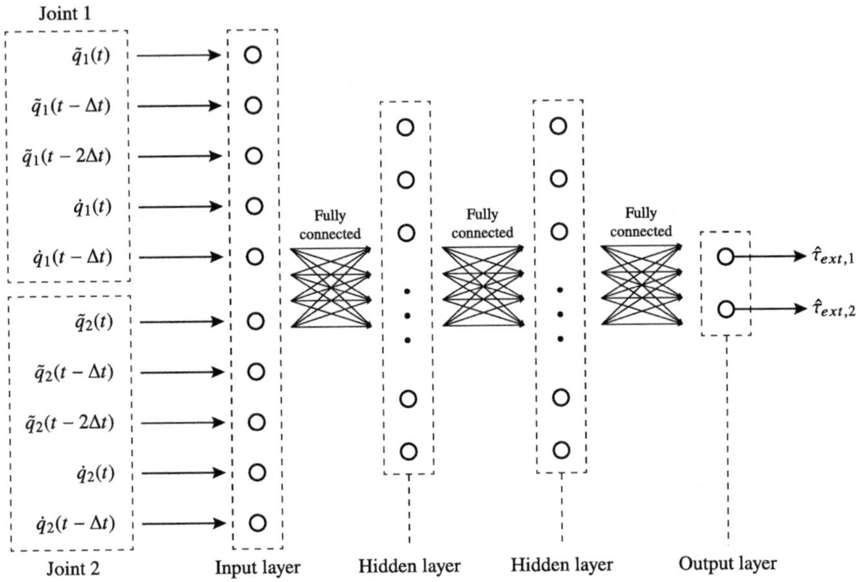

Fig. 4.2 The fully-connected neural network architecture proposed in [4]

4.2.2.1 Estimating the External Joint Torque

In the approach of estimating the external joint torque with learning, all or part of the dynamics model which originally estimates the external joint torque is replaced with a neural network. This approach has the advantage of not requiring accurate identification of the model parameters, and can also identify the collided link since the external torques are estimated for each joint.

In [4], a multilayer FNN is trained to estimate the external joint torque τ_{ext} without using the dynamics model, where only the joint signals are used as the network input. The robot motions the authors test for are two-joint planar motions. For each joint, the position $q(t)$ and the velocity $\dot{q}(t)$ are measured. Given the desired joint position $q_d(t)$, the network inputs are

$$\{\tilde{q}_i(t), \tilde{q}_i(t-\Delta t), \tilde{q}_i(t-2\Delta t), \dot{q}_i(t), \dot{q}_i(t-\Delta t)\},\ i = 1, 2, \tag{4.4}$$

where $\tilde{q}_i(t)$ are the current joint position errors

$$\tilde{q}_i(t) = q_{d,i}(t) - q_i(t),\ i = 1, 2 \tag{4.5}$$

and $\tilde{q}_i(t-\Delta t), \tilde{q}_i(t-2\Delta t)$, and $\dot{q}_i(t-\Delta t)$ are the past joint measurements for each joint. As illustrated in Fig. 4.2, the FNN (after it is trained) takes the above measurements (4.4) as input and computes the estimated external joint torques for each joint $\hat{\tau}_{ext,i}$, which are later thresholded component-wisely for collision detection. To show

Table 4.1 The joint signals used as the network input in [6]

Joint signal	
Joint position	q
Joint velocity	$\dot{q}$
Reference joint velocity	$\dot{q}_r$
Reference joint acceleration	$\ddot{q}_r$
Joint position error	$\tilde{q}$
Joint velocity error	$\dot{\tilde{q}}$
Active motor torque	τ_m
Torque from inverse dynamics	τ_{inv}
Gravitational torque	τ_{grav}
Reference torque	τ_{ref}
Estimated external joint torque	$\hat{\tau}_{ext}$

the generalization capability of the trained network, the authors test the network for configurations that are outside the joint position range for which the network was trained; the external joint torque is accurately estimated even outside the range.

In [5], which is another work conducted by the same authors, the two-joint planar robot motions are extended to three-joint spatial motions, but this time with the help of joint torque sensors and also the gravity vector of the dynamics model. The network inputs are

$$\{\tilde{q}_i(t), \tilde{q}_i(t-\Delta t), \dot{q}_i(t), \tau_{msr,i}(t), g_i(t)\}, \; i = 1, 2, 3, \tag{4.6}$$

where $\tau_{msr,i}(t)$ are the measured joint torques and $g_i(t)$ is the ith component of the dynamics gravity vector $g(q)$. This time, the FNN which uses only one hidden layer computes three outputs, where the outputs are the estimated external joint torques of the three joints, $\hat{\tau}_{ext,i}$, $i = 1, 2, 3$. The generalization capability of the trained network is analyzed similarly as done in [4].

4.2.2.2 End-To-End Networks

The end-to-end networks directly output the predicted collision index $\{True, False\}$ without the need for a thresholding process. In [6], the authors use a 1-D CNN to learn the underlying relation between a time-stacked set of joint signals and the true collision index. The true collision index, which is used as the true label during training, is acquired by an FSR attached to a 6-DOF collaborative robot manipulator. A total of 11 joint signals are used together for the network input, which is composed of both model-free joint signals such as $q(t)$, $\dot{q}(t)$, $\tilde{q}(t)$, and $\tau_m(t)$, and model-based signals including the estimated external joint torque $\hat{\tau}_{ext}(t)$. The full list of the 11 input signals is organized in Table 4.1. We refer the reader to [6] for a detailed description of each signal. The input signals are stacked into

$$x(t) = \left[q^T(t), \dot{q}^T(t), \ldots, \hat{\tau}_{ext}(t)\right]^T \in \mathbb{R}^{66}, \tag{4.7}$$

and the history of $x(t)$ captured by a moving time window of length W

$$X(t) = [x(t - W + \Delta t), \ldots, x(t - \Delta t), x(t)] \in \mathbb{R}^{W \times 66} \tag{4.8}$$

is given as the 1-D CNN input.

One important assumption the authors make is that the robot performs only cyclic (repetitive) motions. With this assumption, the cyclic normalization technique introduced in Sect. 3.5 is applied; the cycle-normalized signal

$$\overline{X}(t) = X(t) - X_{ref} \tag{4.9}$$

is used as the network input for both training and testing, where X_{ref} depends on the executed motion. The trained 1-D CNN is tested for unseen (i.e., not included in the training dataset) cyclic motions and for a different robot platform, validating its motion and robot generalization capabilities. One note we would like to make is that the 1-D CNN architecture proposed in [6] is quite heavy, requiring a computation time of approximately 0.8 ms. Figure 4.3 illustrates the framework of [6] described above.

Another end-to-end approach that is based on an FNN and the Bayesian decision theory is the framework presented in [7]. The proposed framework requires joint torque sensors attached to the joints, where a total of 36 signals are extracted from both time- and frequency domain features of the joint torque sensor measurement; the dynamics model of the robot is not required. The FNN is trained with the 36-dimensional input signals, and an auxiliary online diagnosor designed based on the Bayesian decision theory not only detects the applied external force but also distinguishes accidental collisions from intentional contacts in real-time, together

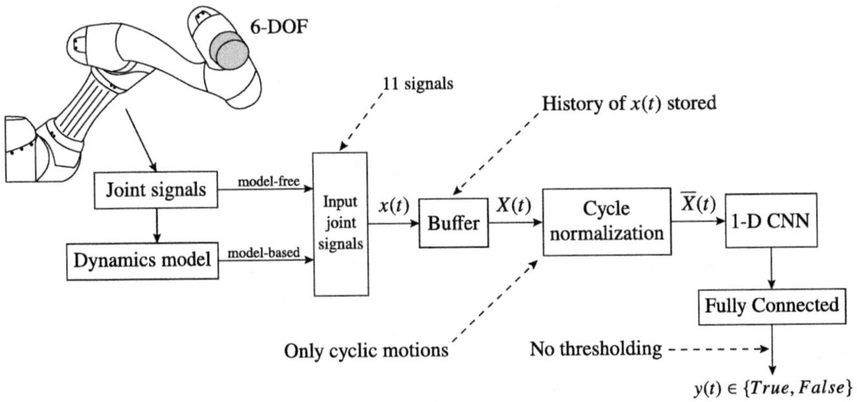

Fig. 4.3 The 1-D CNN-based collision detection framework proposed in [6]

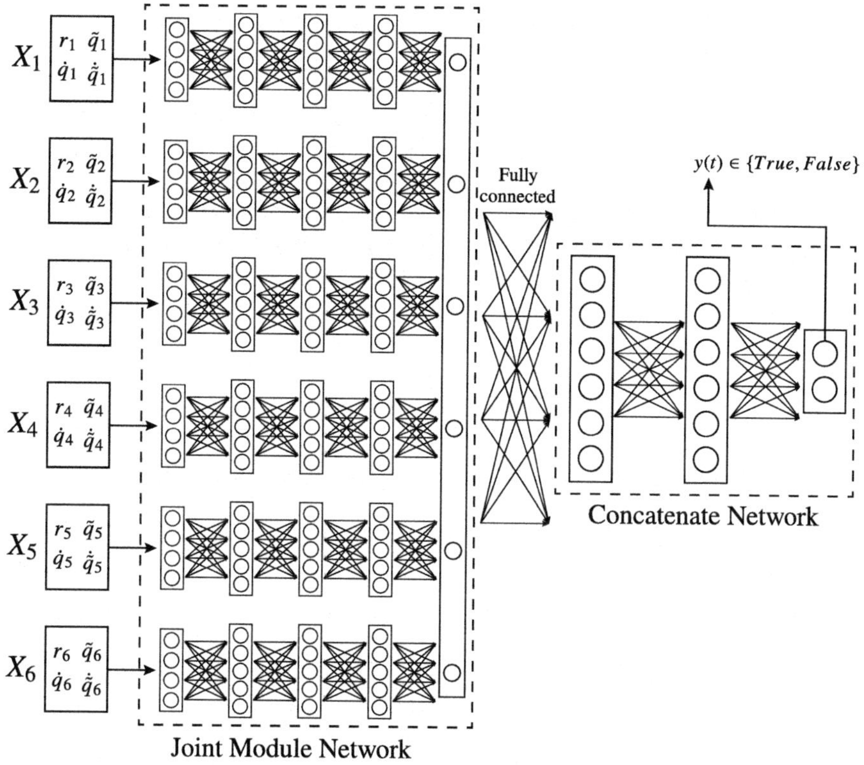

Fig. 4.4 The modularized neural network-based collision detection framework proposed in [8]

providing a confidence index that indicates the output's reliability. The 1-D CNN-based method we have developed also outputs the binary predicted collision index and will be described in detail in later sections.

The last end-to-end method to be introduced is the framework presented in [8], which so far shows the most improved performance among the end-to-end approaches based on the dynamics model. To achieve optimal performance (in terms of training data efficiency, detection performance, and computation), the authors check the detection performance for various network architectures and combinations of input variables (monitoring signals) under limited collision training data. The final structure of the proposed framework is illustrated in Fig. 4.4. The authors propose a modularized neural network (MNN) which is composed of the *joint module network* and the *concatenate network*. The joint module network decouples the joints of a 6-DOF collaborative robot manipulator and extracts features for each joint, and the concatenate network determines whether a collision occurred or not using the features extracted from the joint module network. By decoupling the joints, the parameter search space and computation are reduced compared to an FNN, which leads to sample-efficient training.

The optimal combination of monitoring signals selected as the network input is

$$\{r(t), \dot{q}(t), \tilde{q}(t), \dot{\tilde{q}}(t)\}, \tag{4.10}$$

where r is the output of the momentum observer that does not consider the frictional torque (3.24), and $\tilde{q}$ and $\dot{\tilde{q}}$ are the joint position error (4.5) and velocity error respectively. The joint velocity $\dot{q}$ is chosen so that the network can internally compensate for the frictional joint torque which is included in r. For the error-related factors $\tilde{q}$ and $\dot{\tilde{q}}$, the authors show that the two signals differ distinctively during a collision and under collision-free motions. The monitoring signals of joint i, i =1–6, are stacked into

$$x_i(t) = \left[r_i(t), \dot{q}_i(t), \tilde{q}_i(t), \dot{\tilde{q}}_i(t)\right] \in \mathbb{R}^4, \tag{4.11}$$

and the history of $x_i(t)$ including four previous values are stacked into

$$X_i(t) = [x_i(t - 40\Delta t), x_i(t - 30\Delta t), x_i(t - 20\Delta t), x_i(t - 10\Delta t), x_i(t)] \in \mathbb{R}^{20}. \tag{4.12}$$

Then, the final joint-stacked variable

$$X(t) = [X_1(t), X_2(t), X_3(t), X_4(t), X_5(t), X_6(t)] \in \mathbb{R}^{120} \tag{4.13}$$

is given as the MNN input.

The detection performance between the trained MNN and two existing methods is compared, validating their optimal network architecture and the combination of input variables. The authors also present a method to transfer the network trained for one robot to another robot (with the same model) without performance degradation, which will be described in Chap. 5.

4.2.2.3 Computing Collision Prediction Score

The last group of the supervised learning-based detection methods provides the user with a scalar collision prediction score as the output. The scalar prediction score increases when an external force is applied and stays near zero otherwise. The score is then thresholded with a constant scalar-valued threshold to be transformed into the predicted collision index, therefore providing an easy and intuitive sensitivity adjustment. To the best of our knowledge, the SVMR-based method we have developed is the only method that belongs to this group. The details are described in later sections.

4.2.3 *Summary of the Supervised Learning-Based Detection Methods*

In this section, we summarize the key features of the supervised learning-based detection methods (the algorithms described above and the two methods we have developed). Table 4.2 shows the summarized features of the supervised learning-based detection methods.

4.3 Robot Collision Data

This section describes the robot collision dataset used for training and validating the supervised learning-based detection methods described in the next section (and also for the unsupervised learning-based methods described in Chap. 5). Data collection is performed with a 6-DOF collaborative robot manipulator (Doosan Robotics, *M0609*: rated payload 6 kg, weight 27 kg).

4.3.1 *True Collision Index Labeling*

As described in Chap. 2, the true collision index which is used as the true label for training and testing the detection methods is acquired by an FSR. With the collision index labeling circuit using an FSR (Fig. 4.5a) and the collision index labeling tool with the FSR integrated into it (Fig. 4.5b), the binary true collision index (*True* for collision and *False* for no collision) is collected for force samples larger than 3 N (Fig. 4.5c, d). Details are described in Sect. 2.3.1.

4.3.2 *Collision Scenarios*

We collect the robot collision dataset for three collision scenarios: (1) hard collisions, (2) soft collisions, and (3) collision-free motions. The reason for separately collecting the data of hard and soft collisions is to obtain a collision dataset for as diverse collision situations as possible. The collisions are applied randomly on links 2 through 6 as shown in Fig. 4.6 with the collision index labeling tool, in all directions relative to the direction in which the collided link moves; the link is hit/pushed, e.g., on the front (link velocity and force are in opposite directions), on the back (link velocity and force are in the same direction), or even on the side. For both collecting the collision and collision-free data, the robot is programmed to execute random point-to-point six-joint motions. While collecting the data, we attach three types of payloads to the end-effector of the robot manipulator for each of the three collision scenarios: (1) no payload, (2) 3.3 kg payload, and (3) 5.0 kg payload.

Table 4.2 Supervised learning-based collision detection methods for robot manipulators (JTS: joint torque sensor, #J: the number of joints, C: collision, CF: collision-free)

Methods	Xu [2]	Lim [3]	Sharkawy [4]	Sharkawy [5]	Heo [6]	Zhang [7]	Kim [8]	Park SVMR [1]	Park 1-D CNN [1]
Additional sensor requirement	X	X	X	JTS	X	JTS	X	X	X
Dynamics model requirement	O	O	X	$g(q)$	O	X	O	O	O
Friction compensation	O	O	–	–	O	–	O	O	O
Model uncertainty compensation	O	O	–	–	O	–	O	O	O
Number of detection thresholds	#J	#J	#J	#J	0	0	0	1	0
Learning algorithm (network)	FNN	LSTM	FNN	FNN	1-D CNN	FNN	MNN	SVMR	1-D CNN
Required training data (collision type)	CF	CF	C CF	C CF	C CF	C CF	C CF	C CF	C CF
Generalization capability	–	–	Joint position range		Cyclic motions, robot	–	Unseen payload, robot	Aging of the robot/ sensors, unseen payload	
Validated robot motions	1-joint	2-joint vertical	2-joint planar	3-joint	6-joint cyclic	7-joint fixed initial	6-joint random point-to-point		
Described section	3.4.2	3.4.2 4.2.1	4.2.2.1		4.2.2.2		3.4.2 4.2.2.2	3.4.2 4.4.1	3.4.2 4.4.2

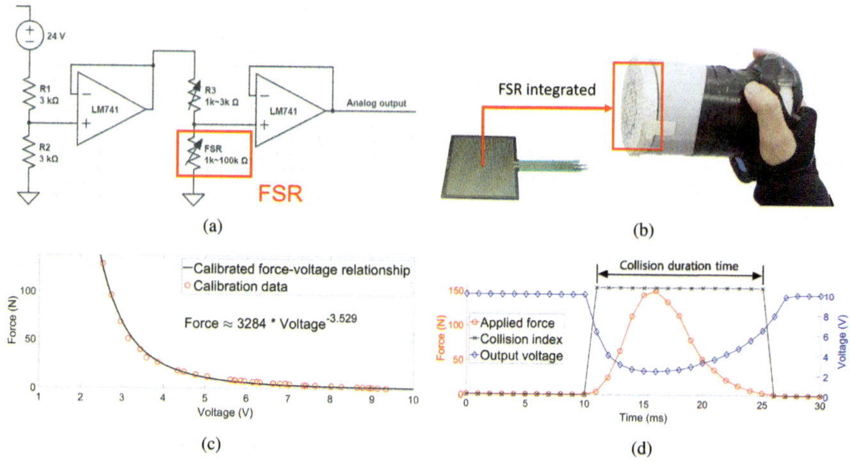

Fig. 4.5 **a** The collision index labeling circuit. **b** The collision index labeling tool. **c** The relationship between the applied force and the circuit output voltage. **d** The applied force during a collision

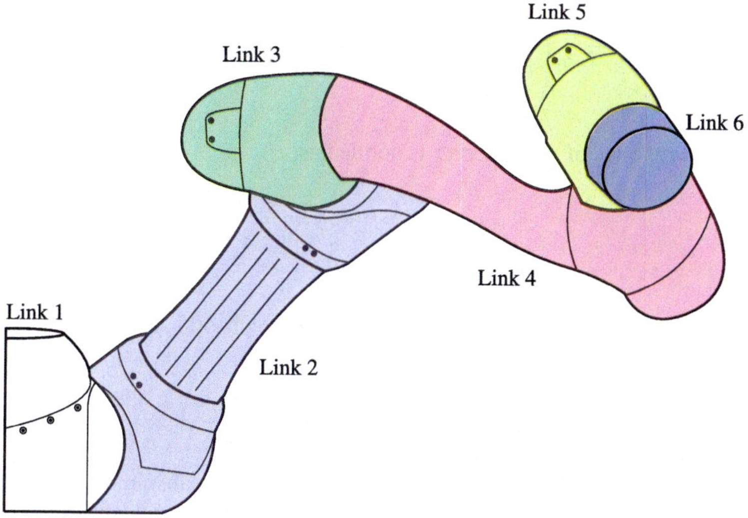

Fig. 4.6 The collision positions of the collision dataset

4.3.3 Monitoring Signal

The monitoring signal used as the input of both the supervised and unsupervised learning-based detection methods is the absolute value of the output of a momentum observer in which the frictional torque τ_F is not considered, given by

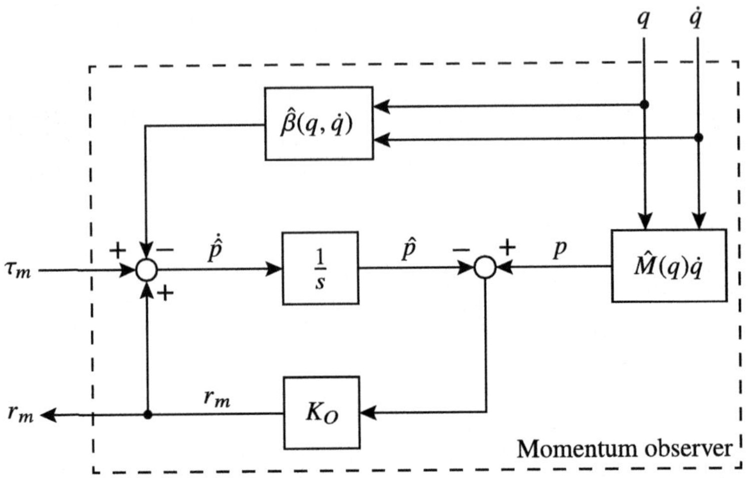

Fig. 4.7 The block diagram of the momentum observer that does not consider the frictional torque

$$r = K_O \left(p(t) - \int_0^t (\tau_m - \hat{\beta} + r) ds - p(0) \right) \in \mathbb{R}^n, \tag{4.14}$$

where the dynamics of the observer is defined as

$$\dot{\hat{p}} = \tau_m - \hat{\beta}(q, \dot{q}) + r, \tag{4.15}$$

$$\dot{r} = K_O(\dot{p} - \dot{\hat{p}}). \tag{4.16}$$

Under ideal conditions, the dynamic relation between $r(t)$ and the external joint torque τ_{ext} is

$$\dot{r} = K_O(\tau_{ext} - \tau_F - r), \tag{4.17}$$

so that $r(t)$ acts as an estimate for $\tau_{ext} - \tau_F$. The elements of the $n \times n$ diagonal observer gain matrix K_O are all set to 150. We note that the absolute value symbol in $|r(t)|$ is omitted for simplicity of notation. Figure 4.7 illustrates the block diagram of the momentum observer described above; the frictional torque τ_F is removed from the input of the observer compared to Fig. 3.4. The cycle time of the robot's control loop is $\Delta t = 1$ ms.

4.3.4 Signal Normalization and Sampling

Before training the learning algorithms with the monitored signal, we apply two transformations to the signal. First, the monitored signal r of the training data is normalized to be in the range [0, 1] (unity) with the following normalization process

$$r_{[0,1]} = \frac{r - r_{\min}}{r_{\max} - r_{\min}}, \tag{4.18}$$

where $r_{\max}$ and $r_{\min}$ are the maximum and minimum values of r computed over the training data, and $r_{[0,1]}$ is the normalized signal used for training. For the supervised learning-based detection methods, since both the data for collision and collision-free motions of the robot are used for training, the maximum and minimum values are computed over both the collision and collision-free data. On the other hand, for the unsupervised learning-based detection methods, which only use the data for collision-free motions for training, the maximum and minimum values are computed only over the collision-free data. The monitored signal of the test data is also transformed by the normalization process described above, using the same $r_{\max}$ and $r_{\min}$ computed over the training data. For the remainder of this book, the normalization symbol [0, 1] in $r_{[0,1]}$ is omitted for simplicity of notation.

Second, the normalized signal $r(t) \in \mathbb{R}^6$ ($\Delta t = 1$ ms) is sampled over a sliding time window of size $t_W + 1$ at a sampling interval of t_I from the data buffer

$$R_{bf} = [r(t - t_W), \ldots, r(t)] \in \mathbb{R}^{6\times(t_W+1)}, \tag{4.19}$$

so that $N + 1$ measurements (time instances) are sampled, where $N = t_W/t_I \in \mathbb{N}$. The sampled measurements are then concatenated into the matrix

$$R = \left[r_{t_{N+1}}, \ldots, r_{t_1}\right] \in \mathbb{R}^{6\times(N+1)} \tag{4.20}$$

as shown in Fig. 4.8. Sampled measurement $r_{t_1} \in \mathbb{R}^6$ is the observation at time t, the most recent observation, while $r_{t_{N+1}} \in \mathbb{R}^6$ is the observation at time $t - t_W$, the most past observation. The specific values of the time sampling parameters, t_W and t_I, are described in the following chapters.

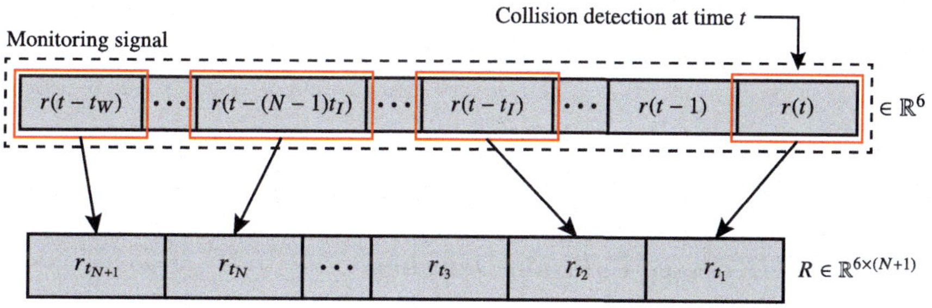

Fig. 4.8 Sampling of the monitored signal $r(t)$ ($\Delta t = 1$ ms). Over the sliding time window of size t_W and at a sampling interval of t_I, r is sampled and transformed into R

Table 4.3 The number of applied test collisions and execution time of collision-free motions for each payload

Payloads	0.0 kg	3.3 kg	5.0 kg	Total
Hard collisions	204	195	179	578
Soft collisions	94	66	49	209
Collision-free (min)	20.6	21.1	20.7	62.4

Table 4.4 The distribution of collision positions of the test collisions

Collided link	Link 2	Link 3	Link 4	Link 5	Link 6	Total
Hard collisions	36	66	310	133	33	578
Soft collisions	10	31	113	37	18	209
Total	46	97	423	170	51	787

4.3.5 Test Data for Detection Performance Verification

The test dataset for detection performance verification consists of all three collision scenarios: hard collisions, soft collisions, and collision-free. Table 4.3 shows the number of test collisions and the execution times of collision-free motions for each payload; a total of 787 collisions (95,681 collision-labeled samples) and 62.4 min of collision-free motions (3,743,905 collision-free-labeled samples) are tested. Table 4.4 and Fig. 4.6 show the distribution of collision positions of the test collisions; total 644 collisions are applied on links 4 through 6, which are more likely to cause collisions with humans than links 2 and 3, in which 143 collisions are applied. Figure 4.9 shows the distribution of collision duration times of the test collisions; hard collisions have collision duration times in the range of 10–100 ms, while soft collisions have collision duration times on the order of 100–700 ms. The peak contact forces of hard collisions range from 4.1 to 515.5 N while soft collisions have 22.4–537.6 N of peak contact force range.

4.4 Collision Detection Using SVMR and 1-D CNN

4.4.1 SVMR-Based Collision Detection

We begin with the SVMR-based model-compensating collision detection algorithm (MCD-SVMR), which can detect both hard and soft collisions integrally with one SVMR model by training the model with both hard and soft collision data (and also collision-free data).

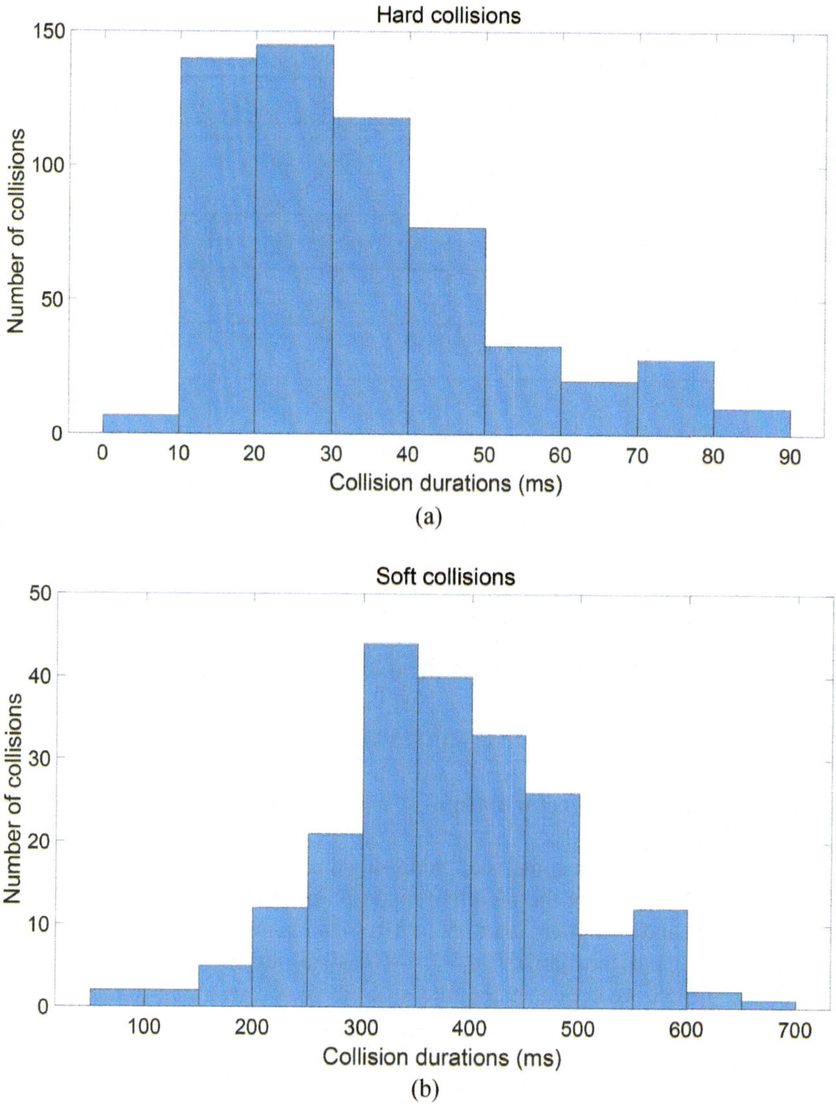

Fig. 4.9 The distribution of collision duration times of the test **a** hard collisions and **b** soft collisions

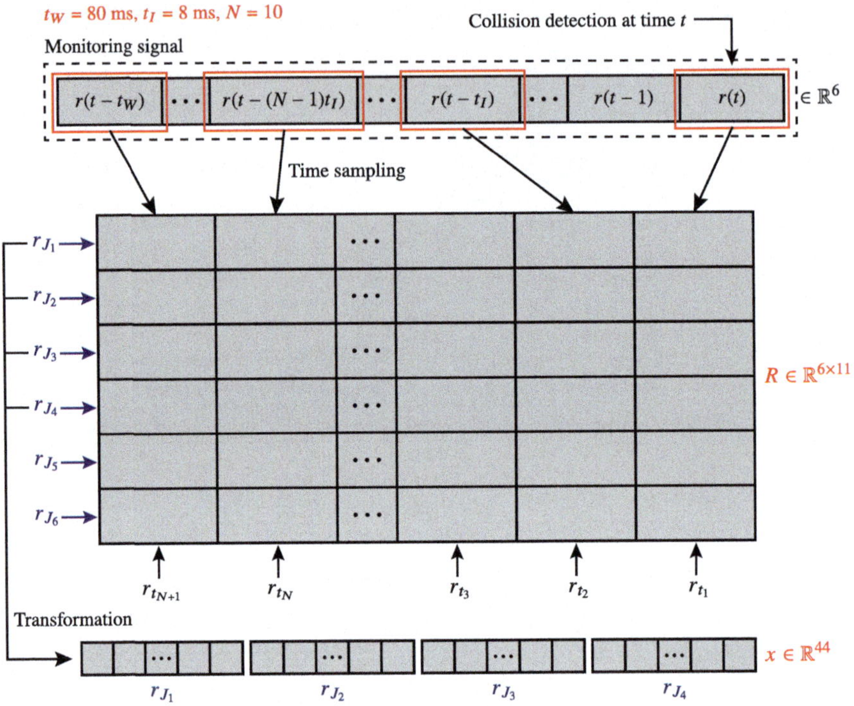

Fig. 4.10 The input feature vector $x(t)$ of MCD-SVMR

4.4.1.1 Input Feature Vector Design

Input feature vector design consists of modifying the monitored signal $r(t) \in \mathbb{R}^6$ into a vector form and labeling the modified feature. For time sampling (Fig. 4.10), we set $t_W = 80$ ms and $t_I = 8$ ms (then, $N = 10$). After sampling the monitored signal and transforming into the matrix $R \in \mathbb{R}^{6\times(N+1)}$ as outlined in Sect. 4.3.4 (4.20), rows 1, 2, 3, and 4 of R are concatenated into a single vector, resulting in the following feature vector $x(t)$:

$$x = \left[r_{J_1}, r_{J_2}, r_{J_3}, r_{J_4}\right]^T \in \mathbb{R}^{4(N+1)}, \tag{4.21}$$

where $r_{J_i} \in \mathbb{R}^{1\times(N+1)}$ denotes the ith row of R. Note that the ith row of R indicates the sampled momentum observer output of joint i. Compared to using the signals of all the joints, the dimension of this feature vector x is smaller, reducing the computation burden for MCD-SVMR. The process described above is illustrated in Fig. 4.10.

The above feature vectors require proper binary labeling of collision (*True*) and collision-free samples (*False*). Figure 4.11 illustrates the feature labeling method for training the SVMR model. We define parameters $t_L (< t_W)$ and m so that a given feature $x(t)$ is labeled as a collision if there are no true collision indices in the range $[t - t_W, t - t_L]$ of the monitored signal r (the dark region of $[r(t - t_W), r(t)]$ in Fig. 4.11) and there are more than m true collision indices in the range $[t - t_L, t]$ of r

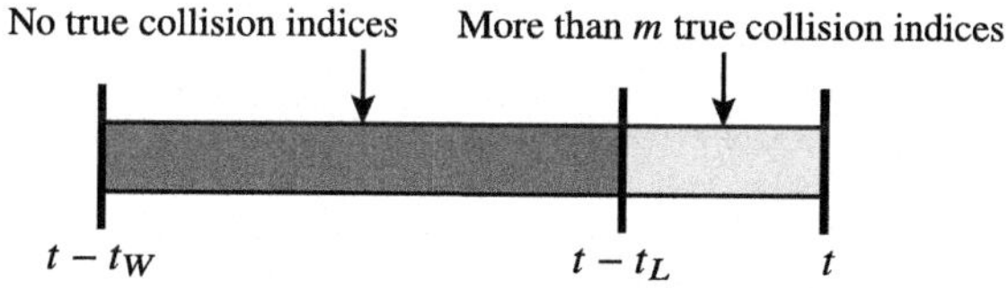

Fig. 4.11 The SVMR feature labeling method

(the light side of $[r(t - t_W), r(t)]$ in Fig. 4.11). We set $t_L = 79$ ms and $m = 3$. This labeling method enables fast detection by learning only the features immediately following collisions.

4.4.1.2 SVMR Training

Since exponential function requires more computation than polynomial functions, we use the squared rational quadratic (SRQ) kernel function [9] instead of the more commonly used radial basis function (RBF) kernel:

$$K_{SRQ}(x, z) = \frac{1}{\left(1 + \frac{\|x-z\|^2}{\sigma^2}\right)^2}. \tag{4.22}$$

We set the hyperparameters of the SVMR model as $C = 1$, $\varepsilon = 0.02$ in (2.27) and $\sigma = 3$ in (4.22). The SVMR model is trained with 1,890,105 feature vectors composed of 105 collision-labeled vectors (12 from hard collision data and 23 from soft collision data, for each of the payloads) and 1,890,000 collision-free-labeled vectors (630,000 for each of the payloads), by using sequential minimal optimization.

4.4.1.3 Collision Detection with the Trained SVMR

The output value $f(x)$ of the trained SVMR model is a scalar value that ranges from values near zero to values near one as depicted in Figs. 4.12 and 4.13. The output value is then thresholded and transformed into the binary predicted collision index by taking into account a threshold value $\alpha \in \mathbb{R}$, which can be used for sensitivity adjustment. The collision decision function $cd_{SVMR} : f(x) \rightarrow \{True, False\}$ for MCD-SVMR is defined as

$$cd_{SVMR}(f(x)) = \begin{cases} True & \text{if } f(x) > \alpha \\ False & \text{if } f(x) \leq \alpha \end{cases}, \tag{4.23}$$

where the binary output indicates whether a collision has occurred or not. Algorithm 1 describes the whole process of collision detection using the trained SVMR.

The output value stays near zero when no external force is applied, increases as the estimated external torques of joints 1–4 (4.21) increase, and decreases as the

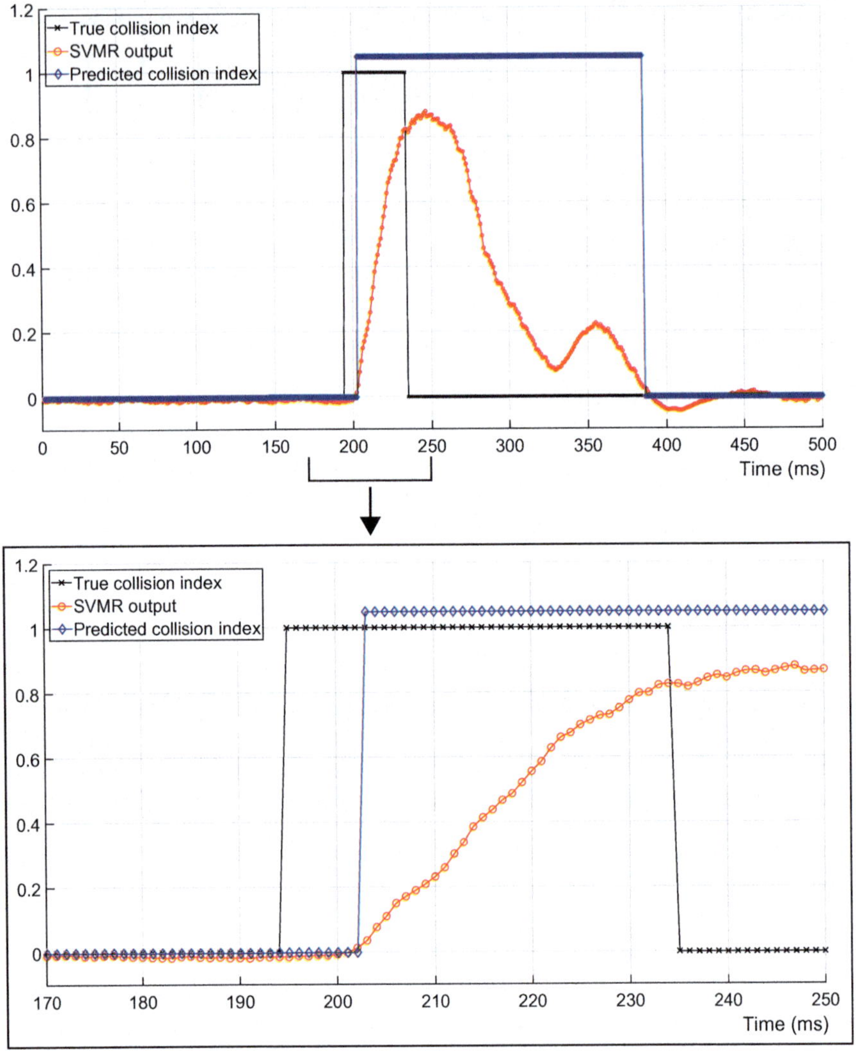

Fig. 4.12 The output scalar value $f(x)$ of the trained SVMR and its corresponding predicted collision index $cd_{SVMR}(f(x))$ when a hard collision occurs

external torques decrease. The larger the estimated external torque and the greater the number of joints affected by the external force (there may exist joints that are not affected by the external force), the greater the SVMR output value. Note that the reason for the SVMR output reaching its peak when the hard collision is already over (Fig. 4.12) is mainly due to the propagation delay, in which the estimated external torque signal responds to the collision in a delayed manner and consequently the SVMR output is also lagged.

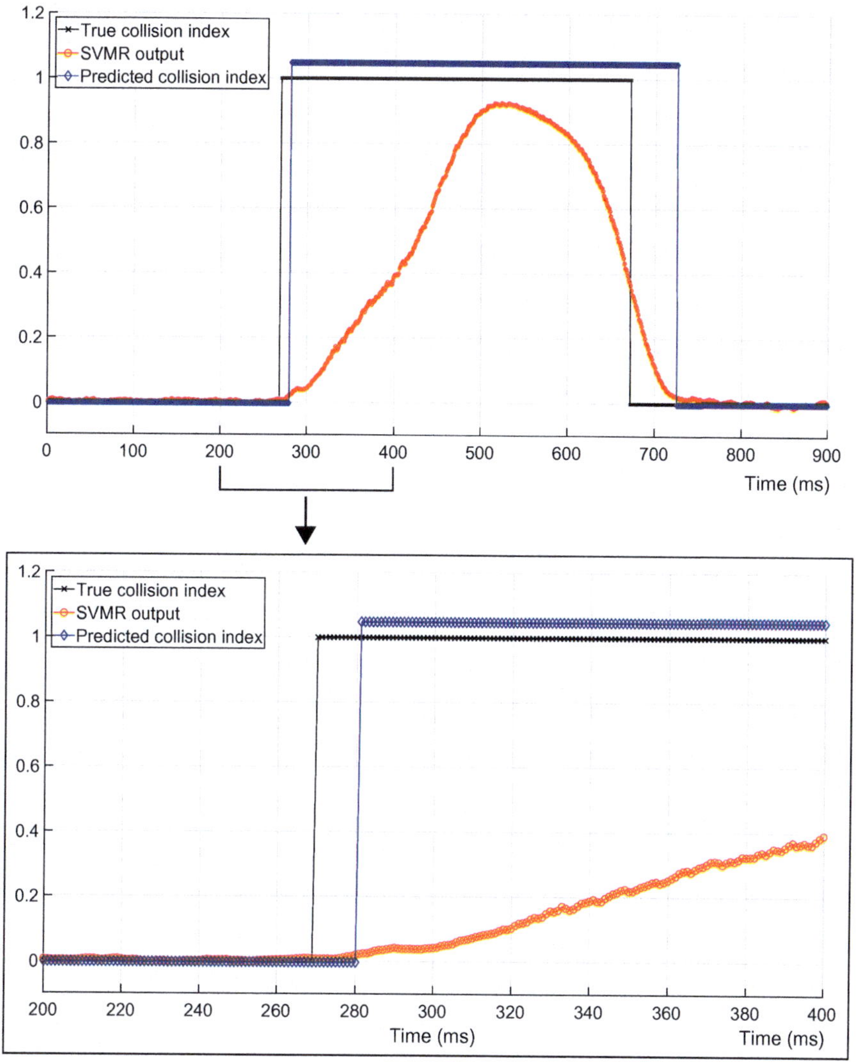

Fig. 4.13 The output scalar value $f(x)$ of the trained SVMR and its corresponding predicted collision index $cd_{SVMR}(f(x))$ when a soft collision occurs

By using SVMR, the detection threshold tuning process is simplified compared to the classical momentum observer-based method, in which the thresholds for each joint have to be determined. The detection sensitivity of the SVMR can be set by adjusting the threshold; high sensitivity is achieved by a low threshold and low sensitivity by a high threshold.

In Appendix, we describe the methodology of expanding MCD-SVMR to the classification of hard and soft collisions using a support vector machine. Although

Algorithm 1 Collision detection with the trained SVMR

$r \leftarrow (0, \ldots, 0) \in \mathbb{R}^6$
$\hat{p} \leftarrow (0, \ldots, 0) \in \mathbb{R}^6$
$cd_{SVMR} \leftarrow False$
while $t > 0$ **do**
  $\hat{M}(q), \hat{C}^T(q, \dot{q}), \hat{g}(q) \leftarrow$ ComputeDynamics$(q, \dot{q})$
  $\hat{p} \leftarrow \hat{p} + (\tau_m - \hat{g}(q) + \hat{C}^T(q, \dot{q})\dot{q} + r) \times \Delta t$
  $r \leftarrow K_O(\hat{M}(q)\dot{q} - \hat{p})$
  $r \leftarrow (|r| - |r|_{min})/(|r|_{max} - |r|_{min})$
  $R_{bf} \leftarrow$ StoreToDataBuffer(r, t_W)
  $R \leftarrow$ SampleMeasurements(R_{bf}, t_I)
  $x \leftarrow$ MakeSVMRInputFeature(R)
  $f(x) \leftarrow b$
  for $i \leftarrow 1$ to l **do**
    $f(x) \leftarrow f(x) + (\alpha_i - \alpha_i^*) \times K_{SRQ}(s_i, x)$
  end for
  if $f(x) > \alpha$ **then**
    $cd_{SVMR} \leftarrow True$
    break
  end if
end while
Reaction to collision

for now, we consider hard and soft collisions in the same manner as just collisions that are potentially dangerous and should be quickly detected, we believe it would be helpful for future research on distinguishing between intended contacts and accidental collisions (the classification phase).

4.4.2 1-D CNN-Based Collision Detection

We now introduce the 1-D CNN-based model-compensating collision detection algorithm (MCD-CNN) that responds to both hard collisions and soft collisions with a single network.

4.4.2.1 Network Input Design

The monitored signal $r(t)$ is sampled over a time window of size $t_W = 50$ ms at sampling interval $t_I = 10$ ms (then, $N = 5$) and transformed into the matrix $R \in \mathbb{R}^{6\times(N+1)}$. The network input feature matrix $Z(t)$ is defined as

$$Z = R^T = \begin{bmatrix} | & | & & | \\ r_{J_1}^T & r_{J_2}^T & \cdots & r_{J_6}^T \\ | & | & & | \end{bmatrix} \in \mathbb{R}^{(N+1)\times 6}, \tag{4.24}$$

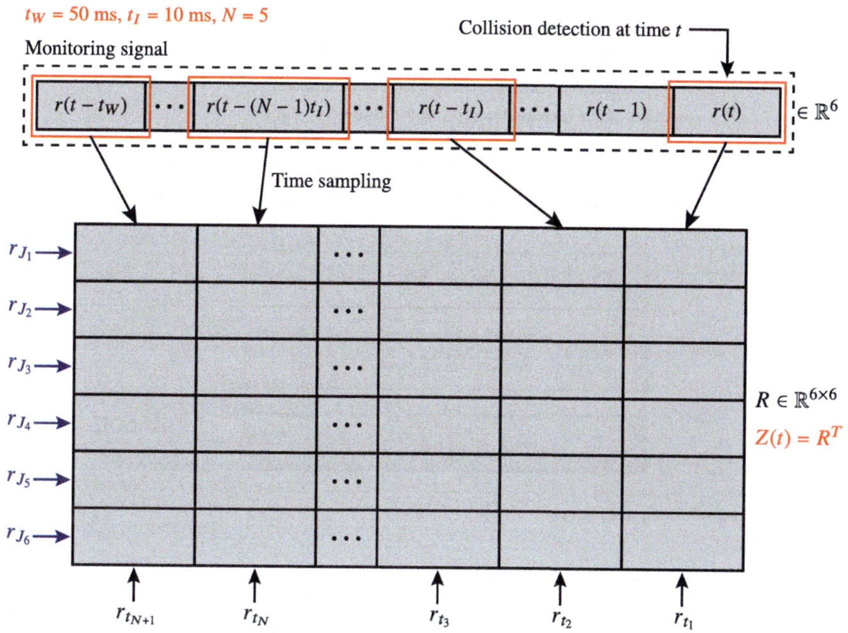

Fig. 4.14 The network input $Z(t)$ of MCD-CNN

Fig. 4.15 The 1-D CNN feature labeling method

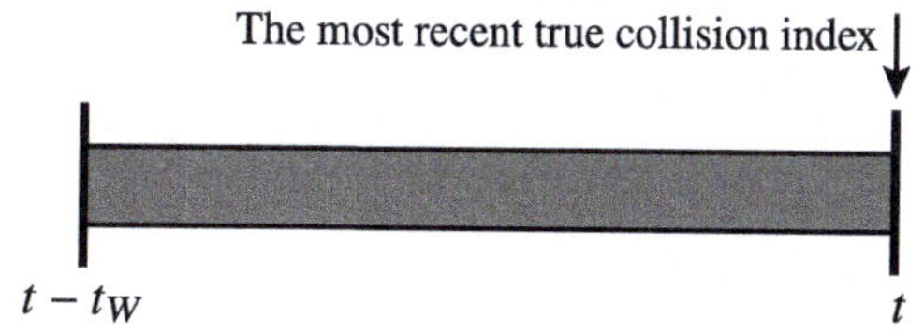

where $r_{J_i} \in \mathbb{R}^{1\times(N+1)}$ denotes the ith row of R. The time sampling process described above is illustrated in Fig. 4.14. A given input feature $Z(t)$ is labeled as collision or collision-free simply by the true collision index of the monitored signal $r(t)$ ($Z(t)$ is labeled as a collision if the true collision index for $r(t)$ is a collision). The labeling method is illustrated in Fig. 4.15.

4.4.2.2 Network Architecture and Training

The MCD-CNN network architecture depicted in Fig. 4.16 contains four 1-D convolution layers, one flatten layer, one fully connected layer, and an output filter. Two of the 1-D convolution layers utilize the same padding, while the other two utilize the valid padding. After passing through the flatten and fully connected layers and the output filter (described in the next section), the binary predicted collision index is output, eliminating the detection threshold tuning process. The ReLU (rectified

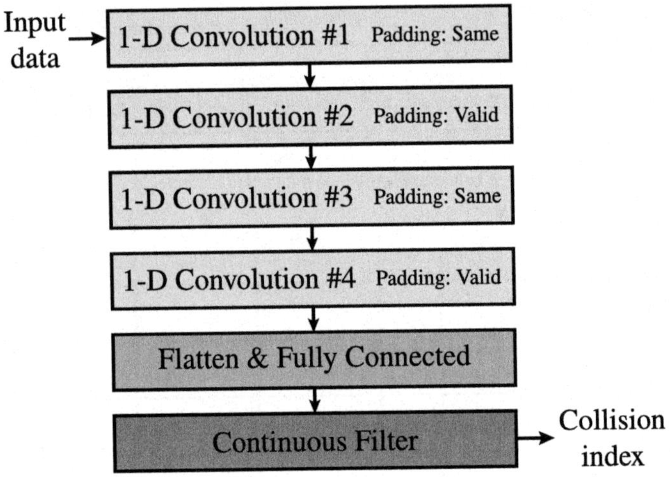

Fig. 4.16 The network architecture of MCD-CNN

Table 4.5 The hyperparameters of the 1-D CNN architecture

Layer	Input (W_{in}, d_{in})	Kernel size k	Stride d_s	Padding	Number of filters	Output (W_{out}, d_{out})
1-D Convolution #1	(6, 6)	3	1	Same	20	(6, 20)
1-D Convolution #2	(6, 20)	3	1	Valid	20	(4, 20)
1-D Convolution #3	(4, 20)	3	1	Same	40	(4, 40)
1-D Convolution #4	(4, 40)	3	1	Valid	40	(2, 40)
Flatten and fully connected	(2, 40)	–	–	–	–	(1)

linear unit) activation function is used in all the layers except the output layer, which uses sigmoid activation. The network is trained with 8,036,355 input features including 16,040 collision-labeled features (no payload—3,422 hard collision, 3,789 soft collision, and 3,695,740 collision-free features, 3.3 kg payload—2,074 hard collision, 2,193 soft collision, and 1,859,525 collision-free features, 5.0 kg payload—2,524 hard collision, 2,038 soft collision, and 2,465,050 collision-free features). The hyperparameters of the 1-D CNN architecture are listed in Table 4.5.

Algorithm 2 Collision detection with the trained 1-D CNN

$r \leftarrow (0, \ldots, 0) \in \mathbb{R}^6$
$\hat{p} \leftarrow (0, \ldots, 0) \in \mathbb{R}^6$
$cd_{CNN} \leftarrow False$
while $t > 0$ **do**
  $\hat{M}(q), \hat{C}^T(q, \dot{q}), \hat{g}(q) \leftarrow \text{ComputeDynamics}(q, \dot{q})$
  $\hat{p} \leftarrow \hat{p} + (\tau_m - \hat{g}(q) + \hat{C}^T(q, \dot{q})\dot{q} + r) \times \Delta t$
  $r \leftarrow K_O(\hat{M}(q)\dot{q} - \hat{p})$
  $r \leftarrow (|r| - |r|_{min})/(|r|_{max} - |r|_{min})$
  $R_{bf} \leftarrow \text{StoreToDataBuffer}(r, t_W)$
  $R \leftarrow \text{SampleMeasurements}(R_{bf}, t_I)$
  $Z \leftarrow \text{MakeCNNInput}(R)$
  $y \leftarrow \text{CNNForward}(Z)$
  if ContinuousFilter $(y, t_c) = True$ **then**
    $cd_{CNN} \leftarrow True$
    break
  end if
end while
Reaction to collision

4.4.2.3 An Output Filtering Method to Reduce False Alarms

Compared to SVMR, 1-D CNN does not require detection threshold tuning and therefore cannot adjust the detection sensitivity. Therefore, we apply a simple continuous output filtering method to reduce false alarms. A collision is signaled only if the output $y \in \{True, False\}$ of the 1-D CNN remains *True* continuously over the interval t_c ms. The output filter eliminates glitches (false positives with duration times in the range 1–t_c ms) in the outputs of the 1-D CNN, making collision detection more robust to disturbances, e.g., tremors in the robot joints. The 1-D CNN architecture introduced earlier is augmented with the continuous filter to complete MCD-CNN as depicted in Fig. 4.16. Algorithm 2 describes the whole process of collision detection using the trained 1-D CNN, where the binary output $cd_{CNN} \in \{True, False\}$ indicates whether a collision has occurred or not.

4.4.3 Collision Detection Performance Analysis

In this section, the performance of the supervised learning-based detection methods is verified for the test dataset described in Sect. 4.3.5—total 787 collisions and 62.4 min of collision-free motions (Tables 4.3 and 4.4). We first evaluate the PRAUC of MCD-SVMR with varying detection thresholds (α) and the PRAUC of MCD-CNN with varying output continuous filter parameters (t_c). Then for each method, with a fixed threshold and a fixed output filter parameter, we measure the average detection delay and the number of detection failures for the collisions. Finally, we measure the average computation times of the algorithms for real-time inference verification.

In Appendix, we describe methodologies of alternative supervised learning-based detection methods: (1) one that uses directly estimated values of the external joint torque as input and (2) a model-free method that uses only the motor current measurements. The experimental results are compared with those of MCD-SVMR and MCD-CNN.

4.4.3.1 Global Performance with Varying Thresholds

For MCD-SVMR, while gradually increasing the detection threshold α from $\alpha = -0.08$ (all labels predicted as *True*) to $\alpha = 0.98$ (all labels predicted as *False*), we calculate the (precision, recall)-pair for each threshold value and plot the precision-recall curve as shown in Fig. 4.17a. Then, the PRAUC is calculated with the precision-recall curve, which can be expressed as a ratio to the maximum possible value 1. A similar procedure is done for MCD-CNN, this time with the output filter parameter t_c varying from $t_c = 1$ ms (no filtering applied) to $t_c = 619$ ms (all the labels predicted as *True* are filtered to *False*). We note that since the continuous filtering can only transform the predicted *True*'s to *False*'s (the opposite is not possible; the existing false negatives cannot be eliminated), the maximum recall is not 1. As illustrated in Fig. 4.17a, b, MCD-SVMR shows a PRAUC of 80.95% while MCD-CNN shows a PRAUC of 82.12%. These results indicate that both supervised learning-based detection methods are well classifying the input features into collision and collision-free for a wide range of detection thresholds and output filter parameters.

4.4.3.2 Detection Delay and Number of Detection Failures

For actual collision detection, the detection threshold α of MCD-SVMR and the output filter parameter t_c of MCD-CNN must be specified. While model-based methods require manual tuning of multiple detection thresholds, MCD-SVMR requires only one constant detection threshold to be tuned while MCD-CNN requires only one constant output filter parameter to be tuned. We set the threshold of MCD-SVMR to be the most sensitive to collisions while not producing any false alarms (the threshold value that corresponds to the largest recall while precision is 1), which can be achieved by setting the threshold to the maximum value of $f(x)$ computed over the collision-free motions. For MCD-CNN, we use the smallest possible output filter parameter value that eliminates all the false alarms, which also corresponds to the largest recall while precision is 1. The threshold is set to $\alpha = 0.0259$ for MCD-SVMR and the filter parameter is set to $t_c = 4$ ms for MCD-CNN.

Table 4.6 shows the detection performance of the supervised learning-based detection methods. MCD-SVMR fails to detect 1 hard collision (0.17% failure rate) and 1 soft collision (0.48% failure rate). MCD-CNN fails to detect 2 soft collisions (0.96% failure rate) and successfully detects all hard collisions. The detection failure rates of both methods are lower than 1%. MCD-SVMR shows an average detection delay

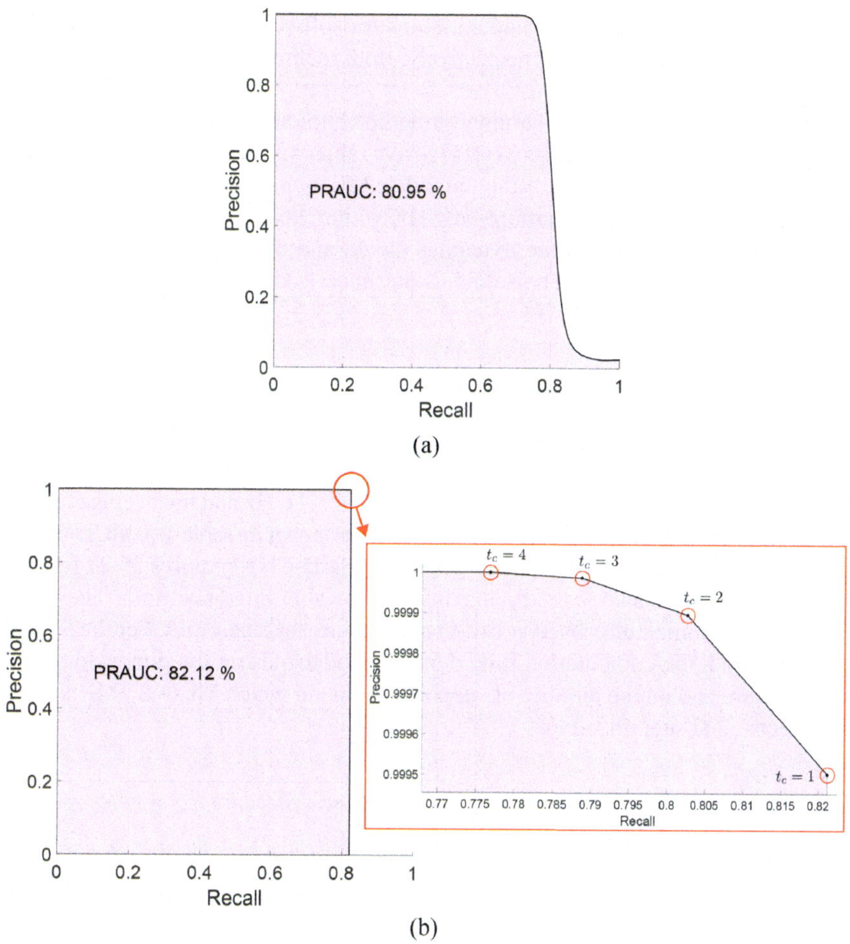

Fig. 4.17 **a** The precision-recall curve with varying thresholds and its PRAUC of MCD-SVMR. **b** The precision-recall curve with varying output filter parameters and its PRAUC of MCD-CNN

Table 4.6 The detection performance of the supervised learning-based detection methods

Methods	MCD-SVMR		MCD-CNN	
Collision type	Hard	Soft	Hard	Soft
Number of detection failures	1/578	1/209	0/578	2/209
Average detection delay (ms)	18.27	32.39	13.14	26.91
Number of false alarms	0/62.4 min		0/62.4 min	
Computation time (μs)	48.61		35.27	

of 18.27 ms for hard collisions and 32.39 ms for soft collisions, while MCD-CNN shows 13.14 ms and 26.91 ms, respectively. Both methods produce no false alarms for about an hour.

The described supervised learning-based detection methods accurately detect a wide range of hard and soft collisions (Fig. 4.9; Tables 4.3 and 4.4) while not producing any false alarms, in spite of the unmodeled frictional torque in the joints. This implies that the supervised learning-based detection methods successfully compensate for the uncertainties in the dynamics model and the unmodeled effects (e.g., friction, sensor measurement noise) by using supervised learning algorithms.

4.4.3.3 Real-Time Inference

For real-time operation, all aspects of the robot operating system including collision detection must run within 1 ms (1 kHz). The detection algorithms are implemented in C++ on a PC with an Intel Core i7-4770 CPU at 3.4 GHz and the average computation times are checked over 10,000 iterations. As shown in Table 4.6, MCD-SVMR requires a computation time of 48.61 μs, while MCD-CNN requires 35.27 μs, both of which are less than 5% of the maximum allowable inference time. The output filter and the momentum observer (4.14) both require less than 1 μs. For the SVMR-based method, the computation time dominantly depends on the dimension of the input feature and on the number of support vectors, in which MCD-SVMR has 298 44-dimensional support vectors.

4.4.4 Generalization Capability Analysis

In this section, we show the generalization capability of the supervised learning-based detection methods for the following two cases: (1) generalization to small changes in the robot dynamics model and random noise added to the motor current measurements and (2) generalization to an unseen payload (a payload with its mass known that was not attached to the end-effector while training the learning algorithms). In Appendix, we also analyze the generalization capability to large changes in the robot mass.

4.4.4.1 Generalization to Small Perturbations

As the robot manipulator is used for a long period of time, small changes may occur in the dynamics model of the robot due to, for example, changes in friction or gear ratio, or wear and tear of the robot parts. Also, due to the aging of the motor current sensors, the measurement noise of the sensors may increase. To roughly simulate the situation described above, we decrease the mass matrix $M(q)$ by 10%, increase the Coriolis matrix $C(q, \dot{q})$ and the gravity vector $g(q)$ by 10%, and add white Gaussian noise with a signal-to-noise ratio of 20 dB to the motor torque measurement $\tau_m = K_i i_m$.

Table 4.7 The generalization capability of the supervised learning-based detection methods to small perturbations in the dynamics model and sensor measurements

Methods	MCD-SVMR		MCD-CNN	
Collision type	Hard	Soft	Hard	Soft
Change of number of detection failures	1/578 ↓ 1/578	1/209 ↓ 1/209	0/578 ↓ 1/578	2/209 ↓ 2/209
Change of average detection delay (ms)	18.27 ↓ 18.98	32.39 ↓ 33.29	13.14 ↓ 12.99	26.91 ↓ 33.96
Number of false alarms	0/62.4 min		0/62.4 min	

To increase the robustness to false alarms, we increase the detection threshold of MCD-SVMR by 4%, in which the original threshold $\alpha = 0.0259$ is now increased to $\alpha = 0.0269$. Table 4.7 shows the detection performance of the supervised learning-based detection methods when small perturbations are applied to the dynamics model and sensor measurements, which is then compared with the detection performance when no perturbations are applied (Table 4.6). MCD-SVMR shows the same detection failure rate as when no perturbations are applied, for both hard (1 detection failure) and soft (1 detection failure) collisions, but the average detection delays are slightly increased (hard collisions: from 18.27 ms to 18.98 ms, soft collisions: from 32.39 ms to 33.29 ms). Compared to when no perturbations are applied, MCD-CNN now fails to detect 1 hard collision, while showing the same detection failure rate for soft collisions (2 detection failures). The average detection delay is decreased from 13.14 ms to 12.99 ms for hard collisions and is increased from 26.91 to 33.96 ms for soft collisions. Both methods produce no false alarms. The results imply that the supervised learning-based detection methods can generalize to small perturbations applied to the dynamics model and random noise added to the motor current measurements.

4.4.4.2 Generalization to an Unseen Payload

We now show that the supervised learning-based detection methods can generalize to an unseen payload with its mass known. Since we only have data for three types of payloads (0, 3.3, and 5.0 kg) and have already used all types for training the previously described MCD-SVMR and MCD-CNN, we train a new SVMR model and a new 1-D CNN with only payload 0 and 3.3 kg data. The new SVMR model is trained with a total of 1,800,120 feature vectors composed of 120 collision-labeled vectors (20 from hard collision data and 40 from soft collision data, for each of the payloads)

Table 4.8 The generalization capability of MCD-SVMR to an unseen payload (trained only with payload 0 and 3.3 kg data, and tested for all payload data)

Payload	0 kg, 3.3 kg		5.0 kg	
Detection threshold (α_1, α_2)	0.0276		0.0357	
Collision type	Hard	Soft	Hard	Soft
Number of detection failures	1/399	1/160	3/179	0/49
Average detection delay (ms)	19.04	27.93	22.62	40.82
Number of false alarms	0/41.7 min		0/20.7 min	

and 1,800,000 collision-free-labeled vectors (900,000 for each of the payloads). The input feature and kernel function hyperparameters are the same as those of the previous MCD-SVMR. The new 1-D CNN is trained with a total of 4,935,307 input features including 10,942 collision-labeled features (no payload—3,406 hard collision, 3,465 soft collision, and 3,276,022 collision-free features, 3.3 kg payload—2,065 hard collision, 2,006 soft collision, and 1,648,343 collision-free features). The input feature and network architecture hyperparameters are the same as those of the previous MCD-CNN.

Since the new MCD-SVMR is trained only with payload 0 and 3.3 kg data, this time we use two detection threshold values: $\alpha_1 \in \mathbb{R}$ for the trained cases (when there is no payload or 3.3 kg payload attached), and $\alpha_2 \in \mathbb{R}$ for the unseen case (when 5.0 kg payload is attached). α_1 is set to the maximum value of $f(x)$ computed over the collision-free motions when no payload and 3.3 kg payload are attached, and α_2 is set in the same way as α_1, this time with the collision-free motions when 5.0 kg payload is attached. Table 4.8 shows the two detection thresholds $\alpha_1 = 0.0276$ and $\alpha_2 = 0.0357$, and the detection performance of the new MCD-SVMR. For the trained cases, the new MCD-SVMR successfully detects 99.75% of the hard collisions (1 failure out of 399 collisions) with an average detection delay of 19.04 ms and 99.38% of the soft collisions (1 failure out of 160 collisions) with an average detection delay of 27.93 ms. For the unseen case, 98.32% of the hard collisions (3 failures out of 179 collisions) are successfully detected with an average detection delay of 22.62 ms and all the soft collisions are successfully detected with an average detection delay of 40.82 ms. Since α_1 and α_2 are set to the maximum values of $f(x)$ computed over the collision-free motions, false alarms do not occur.

Similar to MCD-SVMR, we use two output continuous filters for the new MCD-CNN, one with $t_c = 3$ ms for the trained cases and the other with $t_c = 6$ ms for the unseen case. The filter parameters are set to be the most sensitive to collisions while not producing any false alarms. Table 4.9 shows the detection performance of the new MCD-CNN. For the trained cases, the new MCD-CNN fails to detect 2 soft collisions (1.25% failure rate, average detection delay 23.65 ms) and successfully

Table 4.9 The generalization capability of MCD-CNN to an unseen payload (trained only with payload 0 and 3.3 kg data, and tested for all payload data)

Payload	0 kg, 3.3 kg		5.0 kg	
Output filter (t_c)	3 ms		6 ms	
Collision type	Hard	Soft	Hard	Soft
Number of detection failures	0/399	2/160	1/179	2/49
Average detection delay (ms)	10.79	23.65	15.64	30.57
Number of false alarms	0/41.7 min		0/20.7 min	

detects all hard collisions (average detection delay 10.79 ms). For the unseen case, the new MCD-CNN fails to detect 1 hard collision (0.56% failure rate, average detection delay 15.64 ms) and 2 soft collisions (4.08% failure rate, average detection delay 30.57 ms).

For both the new MCD-SVMR and the new MCD-CNN, to eliminate all the false positives that occur during the unseen case, a higher detection threshold ($\alpha_1 = 0.0276 \rightarrow \alpha_2 = 0.0357$) or heavier output filtering ($t_c = 3$ ms $\rightarrow t_c = 6$ ms) is required compared to the trained cases. Nevertheless, the supervised learning-based detection methods that are trained only with payload 0 and 3.3 kg data can still accurately detect collisions in the case of payload 5.0 kg data as shown in Tables 4.8 and 4.9, successfully generalizing to an unseen payload.

4.4.5 Discussion

The detection performance of the supervised learning-based detection methods we have developed has been experimentally verified for the test dataset we have collected. Results demonstrate that the described methods can accurately detect a wide range of hard and soft collisions in real-time using a light network, compensating for uncertainties in the model parameters as well as unmodeled effects like friction, measurement noise, backlash, and deformations. Moreover, the methods do not require manual tuning of multiple detection thresholds—the SVMR-based method requires only one constant detection threshold to be tuned while the 1-D CNN-based method requires only one constant output filter parameter to be tuned. Furthermore, the generalization capability to the aging of the robot and motor sensors and to an unseen payload has been verified by a set of simulation experiments using the collected dataset.

The experimental results show that the 1-D CNN-based method detects collisions about 5 ms faster than the SVMR-based method, while the SVMR-based method requires much fewer collision data for training, and the detection sensitivity can be adjusted more flexibly. As a general rule, the 1-D CNN-based method is preferable in the event that abundant training data is available, whereas the SVMR-based method performs better with limited data and is useful when flexible detection sensitivity adjustment is needed.

References

1. Park, K.M., Kim, J., Park, J., Park, F.C.: Learning-based real-time detection of robot collisions without joint torque sensors. IEEE Robot. Autom. Lett. **6**(1), 103–110 (2020)
2. Xu, T., Fan, J., Fang, Q., Zhu, Y., Zhao, J.: A new robot collision detection method: a modified nonlinear disturbance observer based-on neural networks. J. Intell. Fuzzy Syst. **38**(1), 175–186 (2020)
3. Lim, D., Kim, D., Park, J.: Momentum observer-based collision detection using LSTM for model uncertainty learning. In: 2021 IEEE International Conference on Robotics and Automation (ICRA), pp. 4516–4522. IEEE (2021)
4. Sharkawy, A.-N., Koustoumpardis, P.N., Aspragathos, N.: Neural network design for manipulator collision detection based only on the joint position sensors. Robotica **38**(10), 1737–1755 (2020)
5. Sharkawy, A.-N., Koustoumpardis, P.N., Aspragathos, N.: Human–robot collisions detection for safe human–robot interaction using one multi-input–output neural network. Soft Comput. **24**(9), 6687–6719 (2020)
6. Heo, Y.J., Kim, D., Lee, W., Kim, H., Park, J., Chung, W.K.: Collision detection for industrial collaborative robots: a deep learning approach. IEEE Robot. Autom. Lett. **4**(2), 740–746 (2019)
7. Zhang, Z., Qian, K., Schuller, B.W., Wollherr, D.: An online robot collision detection and identification scheme by supervised learning and Bayesian decision theory. IEEE Trans. Autom. Sci. Eng. (2020)
8. Kim, D., Lim, D., Park, J.: Transferable collision detection learning for collaborative manipulator using versatile modularized neural network. IEEE Trans. Robot. (2021)
9. Genton, M.G.: Classes of kernels for machine learning: a statistics perspective. J. Mach. Learn. Res. **2**, 299–312 (2001)

Chapter 5
Enhancing Collision Learning Practicality

5.1 Introduction

Despite achieving satisfactory performance, there are two major limitations of the supervised learning-based methods using collision data for training described in Sect. 4.2.2 [1–6]:

- Collection of robot collision data is required: the robot may be damaged during repeated collisions, and it is moreover impractical to collect collision data for each of the mass-produced industrial robots.
- Only the scenarios that are learned as collisions can be robustly detected: for accurate detection, every possible type of collision that can occur (in our case, two types: hard and soft collisions) must be learned.

If the above limitations are effectively addressed, the practicality of learning-based detection methods would be greatly enhanced, making them easier to be replicated for mass-produced industrial robots.

In this chapter, we first introduce several approaches made to overcome the first limitation, including methods that use unsupervised anomaly detection algorithms (unsupervised learning) and a transfer learning-based scheme. Then, we describe in detail the two unsupervised learning-based detection methods we have developed, one using an OC-SVM and the other based on an autoencoder [7]. For the sake of training, all the methods to be presented require only the collision-free motion data of the (new) robot to be trained, effectively addressing the first limitation. That is, robot collisions can be learned even without actual collisions. This advantage makes these detection methods more practical for mass-produced industrial robots compared to the supervised learning-based methods described in Sect. 4.2.2, avoiding the risk of damaging the robots. We also verify that applying unsupervised learning to collision detection can overcome the second limitation, by comparing the detection performance between our unsupervised learning-based detection methods and two

K. M. Park and F. C. Park, *Collision Detection for Robot Manipulators: Methods and Algorithms*, Springer Tracts in Advanced Robotics 155,
https://doi.org/10.1007/978-3-031-30195-7_5

1-D CNNs, one trained with only hard collisions and the other with only soft collisions. To the best of our knowledge, our study is the first to systematically validate the above advantage of unsupervised learning-based collision detection.

5.2 Learning Robot Collisions Without Actual Collisions

There are mainly three approaches to enhance the practicality of the learning-based collision detection methods. As the reader may have already noticed, the friction learning approach introduced in Sects. 3.4.2 and 4.2.1 [8, 9] is one way. Although these methods do not directly learn the detection ability and only compensate for the model-based disturbance observer, they can be considered one way of enhancing learning-based detection practicality. We note that further study is required for this scheme since the existing methods are limited to one- and two-joint motions yet. The second approach is to use unsupervised anomaly detection (or equivalently, unsupervised learning). As described in Sect. 2.4.3, unsupervised anomaly detection algorithms learn the model of normality from a training dataset composed of normal observations and detect anomalous inputs with the trained model. In the case of collision detection, the normal observations are the collision-free motions and the anomalous inputs are the monitoring signals during collisions. The third and most recently proposed method is based on transfer learning (fine-tuning). Transfer learning is a technique of transferring a part of the source network (a pre-trained network) to the target network with fine-tuning using available (often limited) training data [10], which have been successfully applied in many domains such as fault detection in chemical processes [11, 12], motor fault diagnosis [13, 14], and optical inspection [15, 16]. A collision detection network trained for the source robot (with both collision and collision-free data) can be effectively transferred to the target robot without the need for collision data from the target robot.

5.2.1 Using Unsupervised Anomaly Detection

Although not developed based on a robot manipulator but on a humanoid robot, the authors of [17] present a collision detection scheme using an OC-SVM, where collisions that occur during quadruped walking and stair climbing motions are detected. The OC-SVM model is trained to create a representational model of collision-free motions. The monitoring signals selected as the model input are the zero moment point (ZMP) vector of the robot and the force-torque measurements for each limb (wrists and ankles). For the purpose of unifying representational models for different movements, the error (difference between the desired and measured) values instead of raw measurements are used as the final input.

In [18], a collision detection method for a 6-DOF industrial robot manipulator using a sliding-window convolutional variational autoencoder (SWCVAE), which can effectively handle multivariate time-series data, has been proposed. Their method uses only the joint position q and the motor current measurements i_m as the network input, which is therefore a model-free approach. The measurements are sampled at 33.3 Hz ($\Delta t = 30$ ms) over a sliding time window of length $T = 7500$ ms, resulting in a sample length of 250. Then, the network input at time t is given as

$$X(t) = [x(t - T + \Delta t), \ldots, x(t - \Delta t), x(t)]^T \in \mathbb{R}^{250\times 12}, \tag{5.1}$$

where

$$x(t) = \left[q_1(t), \ldots, q_6(t), i_{m,1}(t), \ldots, i_{m,6}(t)\right]^T \in \mathbb{R}^{12}. \tag{5.2}$$

The CNN-structured encoder and decoder use two-dimensional convolution layers to extract features from both axes (time axis and feature axis) of $X(t)$. The output of the autoencoder is the reconstruction probability of $X(t)$, where only the reconstruction probability of the most recent observation $x(t)$ is used to evaluate the anomaly score. Their method is validated for hard collisions that occur during cyclic pick-and-place motions. For performance criteria, the authors use precision, recall, F_1 score, and PRAUC on a per-sample basis, where the F_1 score is defined as the harmonic mean of precision and recall:

$$F_1 = \frac{2 \times Precision \times Recall}{Precision + Recall}. \tag{5.3}$$

We note that detection delay and the number of detection failures and false alarms are evaluated on a per-occurrence basis. Additionally, the trade-off between detection performance and computation time is analyzed with varying sliding time window lengths.

As can be seen from the above studies, we believe that using unsupervised learning can be a promising approach for detecting collisions not only for robot manipulators but also for other types of robots such as humanoids [17], flying robots [19, 20], and mobile robots [21, 22].

5.2.2 *Using Transfer Learning*

In [5], which is an end-to-end method based on an MNN introduced in Sect. 4.2.2.2, a transfer learning method between the source and target robots (with the same model) is presented. The authors use fine-tuning method (updating the weights of the source network with the target robot data) to effectively transfer an MNN trained for two source robots to the target robot without the need for collision data from the target robot, making their framework easier to be replicated for mass-produced industrial robots.

The authors conduct extensive experiments to find the optimal methodology for transferring the source network. The optimal framework can be summarized as fol-

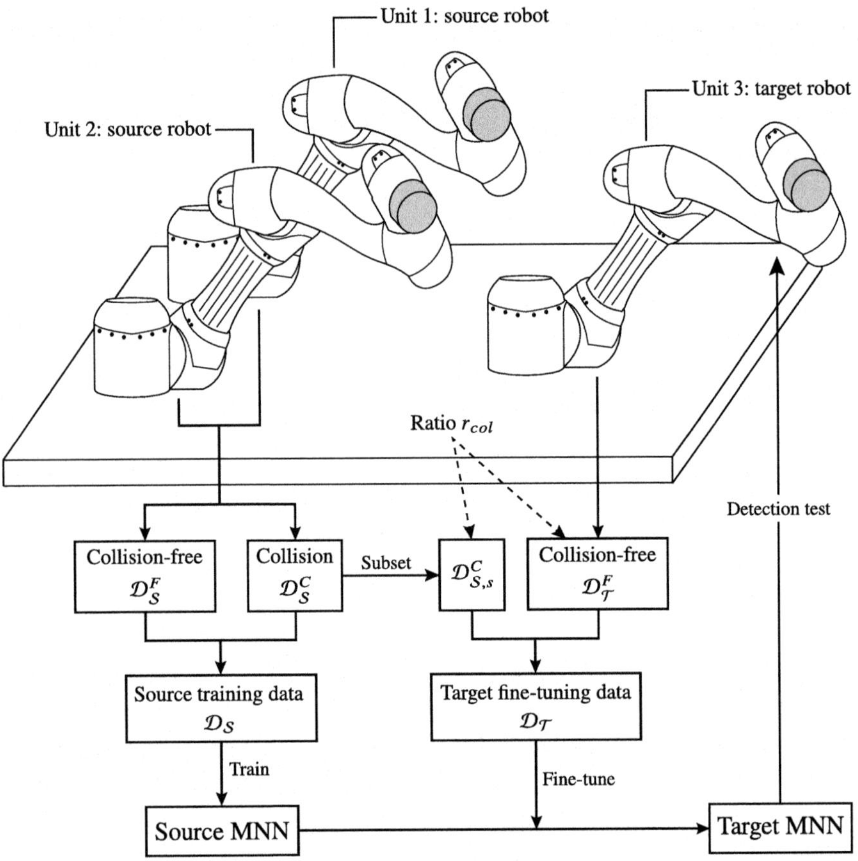

Fig. 5.1 The transfer learning framework proposed in [5]

lows. First, two source robots (unit 1 and unit 2) are used to collect the training data composed of both collision and collision-free motions for training the source network. The training dataset $\mathcal{D}_S$ for the source network can be expressed as

$$\mathcal{D}_S = \{(X_1^S, y_1^S), \dots, (X_{N_S}^S, y_{N_S}^S)\}, \tag{5.4}$$

where X_i^S is the source network input (4.13), $y_i^S \in \{True, False\}$ is the corresponding true collision index, and N_S is the total number of source training data. The source dataset $\mathcal{D}_S$ can be divided into two subsets

$$\mathcal{D}_S^C = \{(X_i^S, y_i^S) \in \mathcal{D}_S | y_i = True\},\ i = 1, \dots, n_S, \tag{5.5}$$

$$\mathcal{D}_S^F = \{(X_j^S, y_j^S) \in \mathcal{D}_S | y_j = False\},\ j = 1, \dots, N_S - n_S, \tag{5.6}$$

where $\mathcal{D}_S^C$ and $\mathcal{D}_S^F$ contain the collision and collision-free samples of $\mathcal{D}_S$ respectively. From the target robot (unit 3), only the collision-free samples $\mathcal{D}_T^F$ are collected:

$$\mathcal{D}_T^F = \{(X_1^T, y_1^T), \ldots, (X_{n_T}^T, y_{n_T}^T)\}, \tag{5.7}$$

where $y_i^T = False$ and n_T is the total number of collision-free fine-tuning (training) data for the target network. To prevent the target network from overfitting to collision-free motions, the authors use a data mixture method of mixing a subset of $\mathcal{D}_S^C$ with $\mathcal{D}_T^F$ for generating the target fine-tuning data $\mathcal{D}_T$. From $\mathcal{D}_S^C$, a subset $\mathcal{D}_{S,s}^C$ with size $n_T \times r_{col}$ is selected, where r_{col} is the ratio between the collision and collision-free samples in $\mathcal{D}_T$. The optimal r_{col} for the final $\mathcal{D}_T$ is set to 2.5%. The pre-trained source network is fine-tuned with the final target dataset

$$\mathcal{D}_T = \mathcal{D}_T^F \cup \mathcal{D}_{S,s}^C. \tag{5.8}$$

Figure 5.1 illustrates the transfer learning process described above.

5.3 Classification of Learning-Based Detection Methods

In this section, we classify the learning-based detection methods for robot manipulators (including [7] which will be described in the following section) into four classes based on full dynamics model requirement and collision training data requirement. Figure 5.2 shows the four groups, with additional information about joint torque sensor requirement, number of detection thresholds, and random motion applicability.

5.4 Collision Detection Using OC-SVM and Autoencoder

In this section, we describe the two unsupervised learning-based detection methods we have developed, one using an OC-SVM and the other based on an autoencoder. Given a training dataset composed of normal (collision-free) motion data only, input feature vectors $\{x_1, \ldots, x_m\}$ are generated by sampling and transforming the dataset. Then by training the learning algorithms with the feature vectors, model functions—$f(x)$ for OC-SVM and $g(x)$ for autoencoder—are acquired in which a test input x^* is considered as an anomaly if $f(x^*) < \beta$ or $g(x^*) > \gamma$, where β and γ are the detection thresholds for each model function. For each method, we describe in detail the design procedure of input feature vectors, the training methodology of the unsupervised learning algorithms, and the collision detection (inference) methodology using the trained OC-SVM and autoencoder.

Then, the detection performance of the unsupervised learning-based collision detection methods is verified for the test dataset described in Sect. 4.3. We first eval-

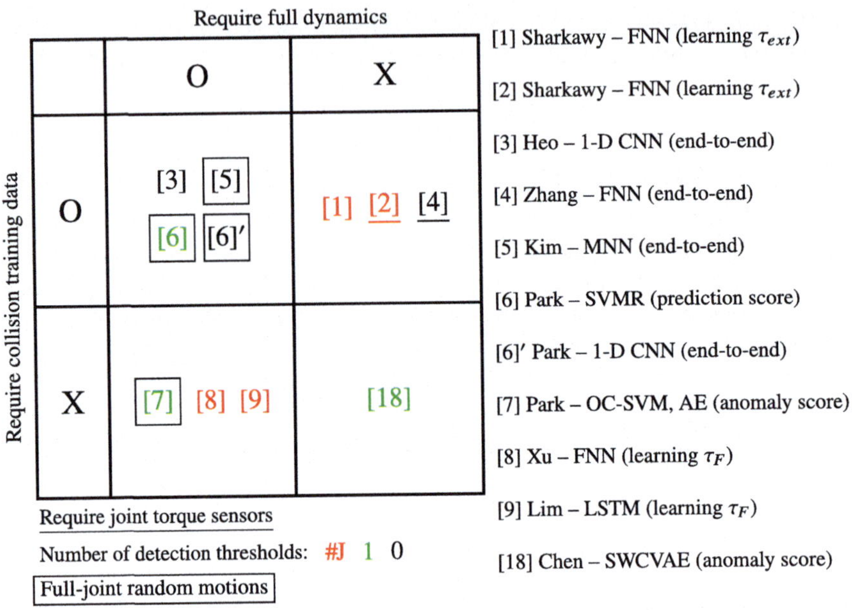

Fig. 5.2 Classification of the learning-based detection methods for robot manipulators

uate the PRAUC with varying detection thresholds (β and γ) for each of the methods. Then with fixed thresholds, we measure the average detection delay and the number of detection failures for the test collisions. Moreover, we compare the detection performance between our unsupervised and supervised learning-based detection methods. Finally, the average computation time and the generalization capability to the aging of the robot and to an unseen payload are analyzed.

5.4.1 OC-SVM-Based Collision Detection

We first describe the OC-SVM-based model-compensating collision detection method (MCD-OCSVM). With one OC-SVM model trained only with collision-free data, MCD-OCSVM is capable of detecting both hard and soft collisions.

5.4.1.1 Input Feature Vector

MCD-OCSVM uses the monitored signal $r(t)$ (which is described in Sect. 4.3) of joints 1–4 as input, in which the computation burden for both training and inference is reduced compared to using signals of all the joints. For time sampling (Fig. 5.3), we set $t_W = 60$ ms and $t_I = 4$ ms (then, $N = 15$). After sampling the monitored signal

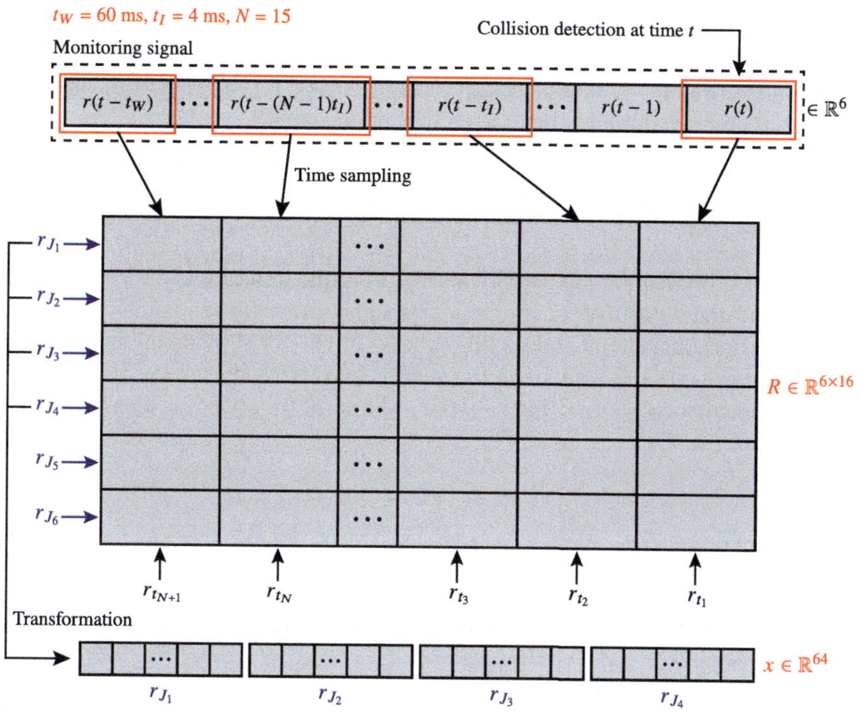

Fig. 5.3 The input feature vector $x(t)$ of MCD-OCSVM

and transforming into the matrix $R \in \mathbb{R}^{6\times(N+1)}$ as outlined in Sect. 4.3.4 (4.20), rows 1, 2, 3, and 4 of R, $r_{J_i} \in \mathbb{R}^{1\times(N+1)}$ $(i = 1, \ldots, 4)$, are concatenated into the input feature vector $x(t)$ as

$$x = \left[r_{J_1}, r_{J_2}, r_{J_3}, r_{J_4}\right]^T \in \mathbb{R}^{4(N+1)}. \tag{5.9}$$

The process described above is illustrated in Fig. 5.3.

Compared to the supervised learning-based detection methods, binary feature labeling of collision and collision-free samples is not required since only the collision-free data is used for training; all the training inputs are labeled as one class, collision-free.

5.4.1.2 OC-SVM Training

For the OC-SVM model, we use the commonly used RBF kernel:

$$K_{RBF}(x, z) = \exp\left(-\frac{\|x - z\|^2}{2\sigma^2}\right), \tag{5.10}$$

where the kernel hyperparameter is set to $\sigma = 0.5$. We set the hyperparameter of the OC-SVM as $\nu = 0.001$ in (2.34). The OC-SVM model is trained with 72,000 collision-free training samples (feature vectors), 24,000 for each payload, where the sequential minimal optimization is used for optimization.

5.4.1.3 Collision Detection with the Trained OC-SVM

As described in Chap. 2, the output value $f(x)$ of the trained OC-SVM model is a scalar value that is positive for normal inputs and negative for anomalous inputs. We modify this original anomaly detection process of OC-SVM by taking into account a threshold value $\beta \in \mathbb{R}$, which can be used for sensitivity adjustment. The collision decision function $cd_{OCSVM} : f(x) \rightarrow \{True, False\}$ for MCD-OCSVM is defined as

$$cd_{OCSVM}(f(x)) = \begin{cases} True & \text{if } f(x) < \beta \\ False & \text{if } f(x) \geq \beta \end{cases}, \tag{5.11}$$

where the binary output indicates whether a collision has occurred or not. Algorithm 1 describes the whole process of collision detection using the trained OC-SVM.

Algorithm 1 Collision detection with the trained OC-SVM

$r \leftarrow (0, \ldots, 0) \in \mathbb{R}^6$
$\hat{p} \leftarrow (0, \ldots, 0) \in \mathbb{R}^6$
$cd_{OCSVM} \leftarrow False$
while $t > 0$ **do**
  $\hat{M}(q), \hat{C}^T(q, \dot{q}), \hat{g}(q) \leftarrow$ ComputeDynamics$(q, \dot{q})$
  $\hat{p} \leftarrow \hat{p} + (\tau_m - \hat{g}(q) + \hat{C}^T(q, \dot{q})\dot{q} + r) \times \Delta t$
  $r \leftarrow K_O(\hat{M}(q)\dot{q} - \hat{p})$
  $r \leftarrow (|r| - |r|_{\min})/(|r|_{\max} - |r|_{\min})$
  $R_{bf} \leftarrow$ StoreToDataBuffer(r, t_W)
  $R \leftarrow$ SampleMeasurements(R_{bf}, t_I)
  $x \leftarrow$ MakeOCSVMInputFeature(R)
  $f(x) \leftarrow -\rho$
  for $i \leftarrow 1$ to l **do**
    $f(x) \leftarrow f(x) + \alpha_i \times K_{RBF}(s_i, x)$
  end for
  if $f(x) < \beta$ **then**
    $cd_{OCSVM} \leftarrow True$
    break
  end if
end while
Reaction to collision

5.4.2 Autoencoder-Based Collision Detection

We now introduce the model-compensating collision detection method using an autoencoder (MCD-AE), in which both hard and soft collisions can be detected with one autoencoder trained only with collision-free data.

5.4.2.1 Network Input and Output

Similar to MCD-OCSVM, the sampled monitored signal $r(t)$ of joints 1–4 are taken as inputs for MCD-AE. The monitored signal $r(t)$ is sampled over a time window of size $t_W = 60$ ms at sampling interval $t_I = 4$ ms (then, $N = 15$) and transformed into the matrix $R \in \mathbb{R}^{6\times(N+1)}$ (Fig. 5.4). We denote the matrix composed of rows 1, 2, 3, and 4 of R as $R_{[1,4]} \in \mathbb{R}^{4\times(N+1)}$. This time, the columns of $R_{[1,4]}$, $r_i \in \mathbb{R}^4$ ($i = 1, \ldots, N+1$), are concatenated into the network input

$$x = \left[r_{N+1}^T, r_N^T, \ldots, r_1^T\right]^T \in \mathbb{R}^{4(N+1)}. \tag{5.12}$$

The procedure described above is illustrated in Fig. 5.4. We denote the reconstructed output of the autoencoder as

$$\hat{x} = \left[\hat{r}_{N+1}^T, \hat{r}_N^T, \ldots, \hat{r}_1^T\right]^T \in \mathbb{R}^{4(N+1)}. \tag{5.13}$$

5.4.2.2 Network Architecture and Training

MCD-AE is composed of five fully connected layers: the input layer (64), three hidden layers (32-6-32), and the output layer (64), where the numbers in parentheses describe the dimensions of each layer. Figure 5.5 depicts the architecture of MCD-AE. The ReLU activation function is utilized in all the layers except the output layer, which uses linear activation. The loss function is set to the commonly used reconstruction error between the input and the reconstructed output (2.37). The autoencoder is trained with 120,000 collision-free training samples, 40,000 for each payload.

5.4.2.3 Collision Detection with the Trained Autoencoder

For collision detection, we first define the anomaly score based on the difference between the original input x and the reconstructed output $\hat{x}$. Since the input signal is time-series data, the original reconstruction error used for training the autoencoder is not the best choice for the anomaly score. As illustrated in Fig. 5.6, when a collision occurs, signals measured more recently within the sliding time win-

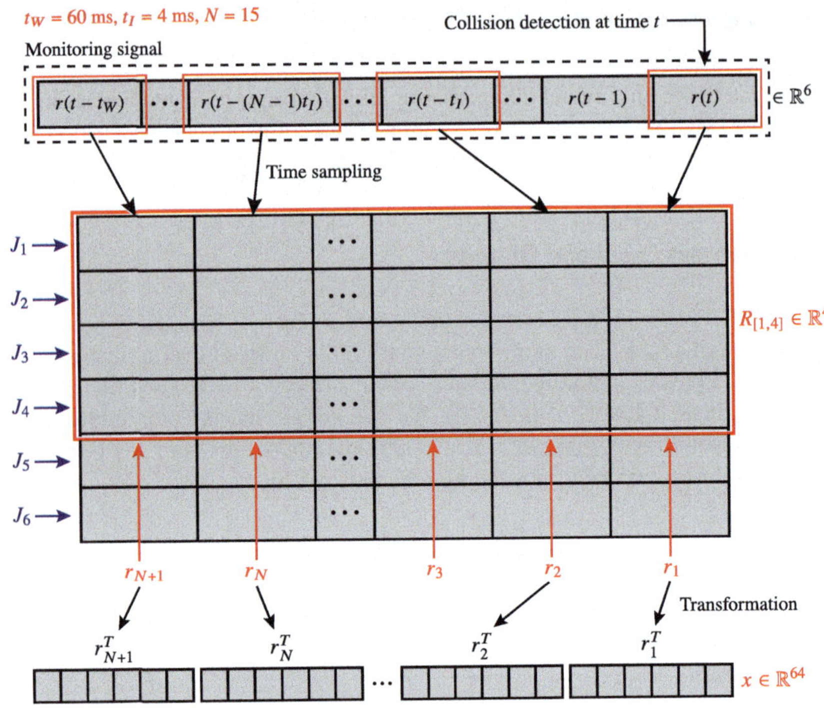

Fig. 5.4 The input feature vector $x(t)$ of MCD-AE

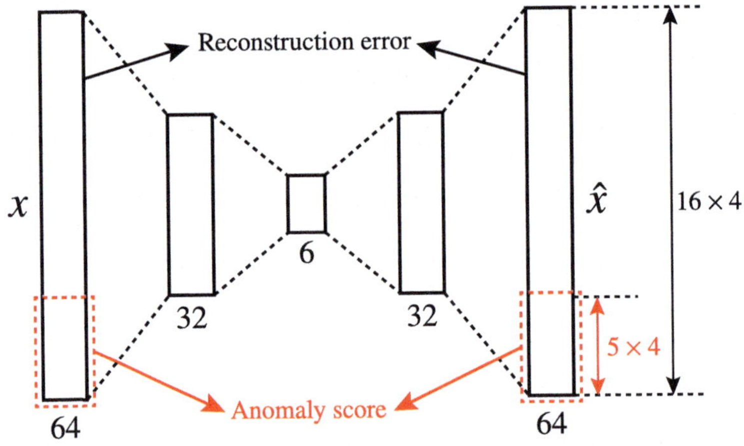

Fig. 5.5 MCD-AE is composed of five fully connected layers. The loss function is set to the reconstruction error between the original input x and the reconstructed output $\hat{x}$, while the anomaly score is calculated with the most recent five observations of x and $\hat{x}$ (areas in red rectangles)

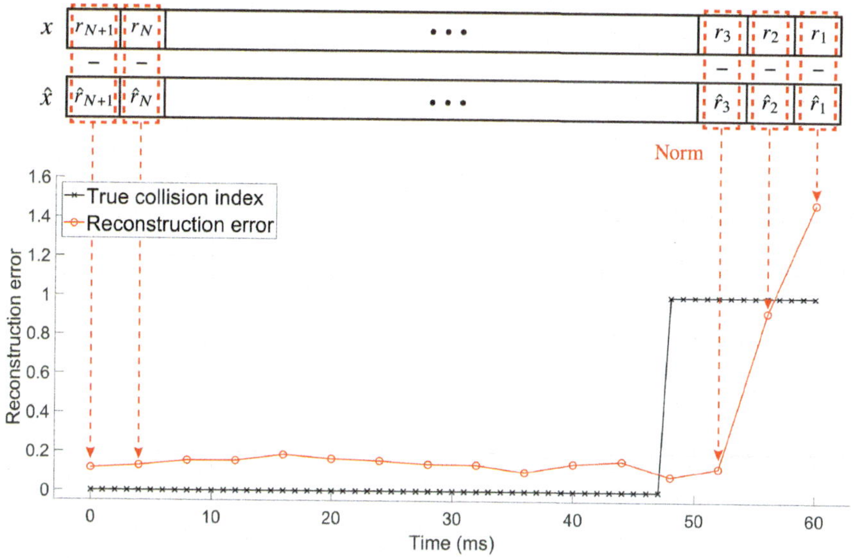

Fig. 5.6 The difference between the original input and the reconstructed output for the sliding time window at $t = 60$ ms. While the reconstruction errors of recent measurements are high, the past measurements have small reconstruction errors

dow $(r_1, r_2, \ldots)$ are affected by the collision earlier than the signals measured in the past $(r_{N+1}, r_N, \ldots)$; while the reconstruction errors of recent measurements $(\left\|r_1 - \hat{r}_1\right\|, \left\|r_2 - \hat{r}_2\right\|, \ldots)$ are high, the past measurements have small reconstruction errors. Therefore, we define the anomaly score $g(x) \in \mathbb{R}$ as

$$g(x) = \left\|\left[r_5^T - \hat{r}_5^T, r_4^T - \hat{r}_4^T, \ldots, r_1^T - \hat{r}_1^T\right]\right\|, \tag{5.14}$$

where the most recent five measurements are used to calculate the anomaly score. The collision decision function $cd_{AE} : g(x) \rightarrow \{True, False\}$ for MCD-AE is defined as

$$cd_{AE}\left(g(x)\right) = \begin{cases} True & \text{if } g(x) > \gamma \\ False & \text{if } g(x) \leq \gamma \end{cases}, \tag{5.15}$$

where $\gamma \in \mathbb{R}$ is the user-defined threshold. Algorithm 2 describes the whole process of collision detection using the trained autoencoder.

Algorithm 2 Collision detection with the trained autoencoder

```
r ← (0, ..., 0) ∈ ℝ^6
p̂ ← (0, ..., 0) ∈ ℝ^6
cd_AE ← False
while t > 0 do
  M̂(q), Ĉ^T(q, q̇), ĝ(q) ← ComputeDynamics(q, q̇)
  p̂ ← p̂ + (τ_m − ĝ(q) + Ĉ^T(q, q̇)q̇ + r) × Δt
  r ← K_O(M̂(q)q̇ − p̂)
  r ← (|r| − |r|_min)/(|r|_max − |r|_min)
  R_bf ← StoreToDataBuffer(r, t_W)
  R ← SampleMeasurements(R_bf, t_I)
  x ← MakeAEInputFeature(R)
  x̂ ← AEReconstruction(x)
  g(x) ← CalculateAnomalyScore(x, x̂)
  if g(x) > γ then
    cd_AE ← True
    break
  end if
end while
Reaction to collision
```

5.4.3 Collision Detection Performance Analysis

In this section, the performance of the unsupervised learning-based detection methods is verified for the test dataset described in Sect. 4.3.5—total 787 collisions and 62.4 min of collision-free motions (Tables 4.3 and 4.4). We first evaluate the PRAUC with varying detection thresholds (β and γ) for each of the methods. Then with fixed thresholds, we measure the average detection delay and the number of detection failures for the collisions. Moreover, we compare the detection performance between the unsupervised and supervised learning-based detection methods. Finally, the average computation times of the methods are measured for real-time inference verification.

5.4.3.1 Global Performance with Varying Thresholds

For MCD-OCSVM, while gradually increasing the detection threshold β from $\beta = -0.71$ (all labels predicted as *False*) to $\beta = 0.58$ (all labels predicted as *True*), we calculate the (precision, recall)-pair for each threshold value and plot the precision-recall curve as shown in Fig. 5.7a. Then, the PRAUC is calculated with the precision-recall curve, which can be expressed as a ratio to the maximum possible value 1. The same procedure is done for MCD-AE, this time with the detection threshold γ varying from $\gamma = 0$ (all labels predicted as *True*) to $\gamma = 4.36$ (all labels predicted as *False*). As illustrated in Fig. 5.7a, b, MCD-OCSVM shows a PRAUC of 90.42% while MCD-AE shows a PRAUC of 90.18%. These results indicate that both unsupervised learning-based detection methods are well classifying the input features into collision and collision-free for a wide range of detection thresholds.

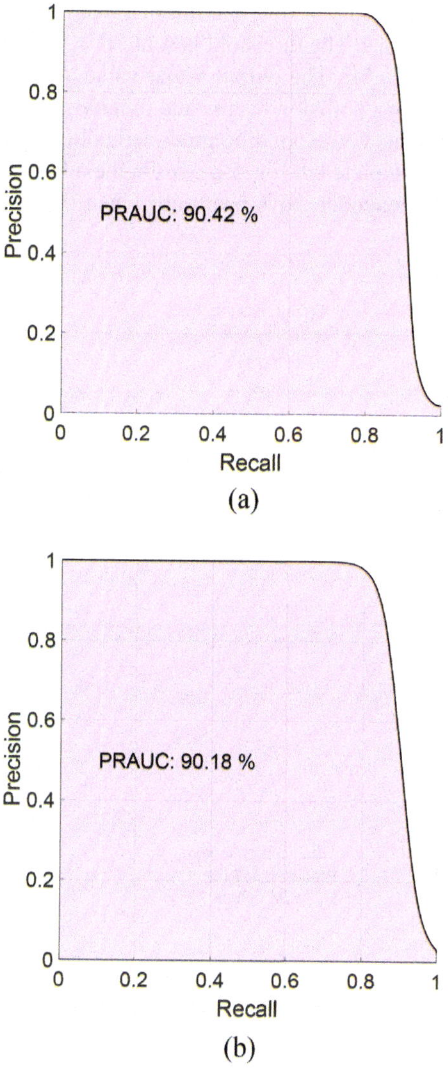

Fig. 5.7 The precision-recall curve with varying thresholds and its PRAUC of **a** MCD-OCSVM and **b** MCD-AE

5.4.3.2 Detection Delay and Number of Detection Failures

For actual collision detection, the detection threshold must be specified, where the detection sensitivity can be adjusted by the threshold. While model-based methods require manual tuning of multiple detection thresholds, the unsupervised learning-based detection methods require only one constant parameter to be tuned. Similar to MCD-SVMR, we set the thresholds to be the most sensitive to collisions while not producing any false alarms (the threshold value that corresponds to the largest recall while precision is 1), which can be achieved by setting β to the minimum value of

$f(x)$ and γ to the maximum value of $g(x)$, both computed over the collision-free motions. The threshold is set to $\beta = -0.382$ for MCD-OCSVM and $\gamma = 0.512$ for MCD-AE. The output scalar value $f(x)$ of the trained OC-SVM, which decreases when a collision occurs and otherwise stays positive or at negative values near zero, and its corresponding predicted collision index $cd_{OCSVM}(f(x))$ of MCD-OCSVM are depicted in Fig. 5.8a, b. On the other hand, the anomaly score $g(x)$ of the trained autoencoder stays near zero when the robot is moving freely and increases when

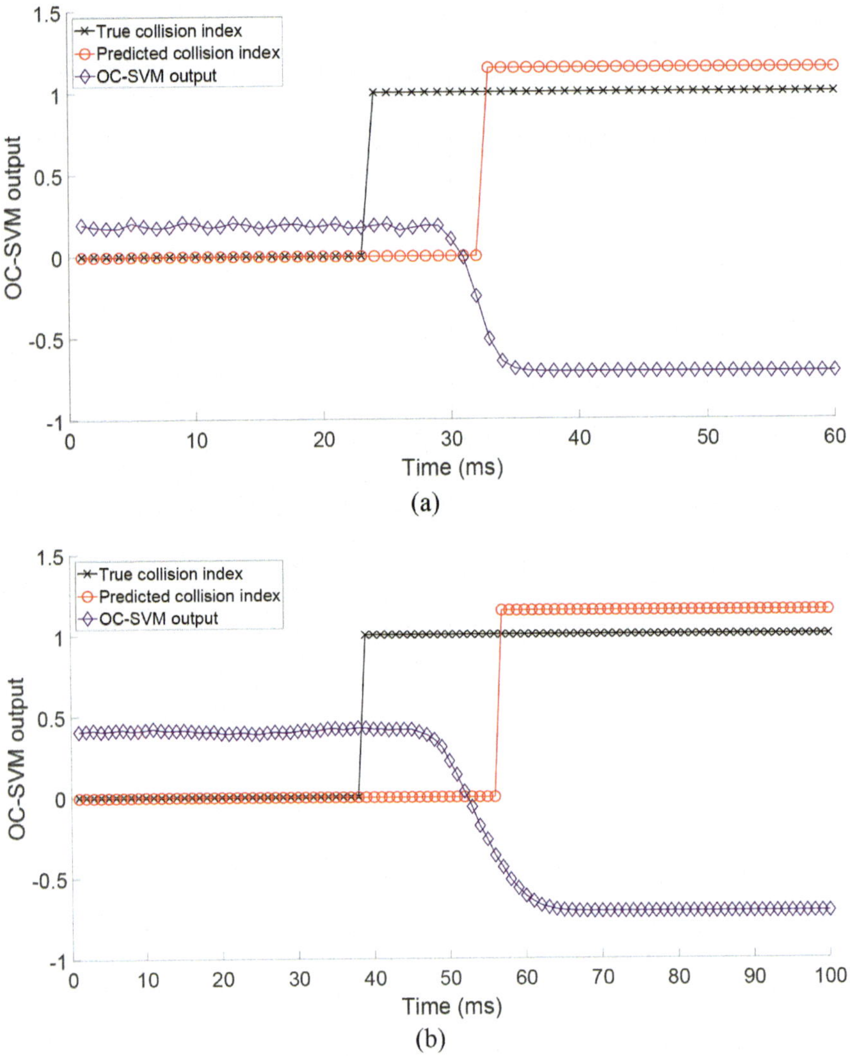

Fig. 5.8 MCD-OCSVM detection results of **a** a hard collision and **b** a soft collision

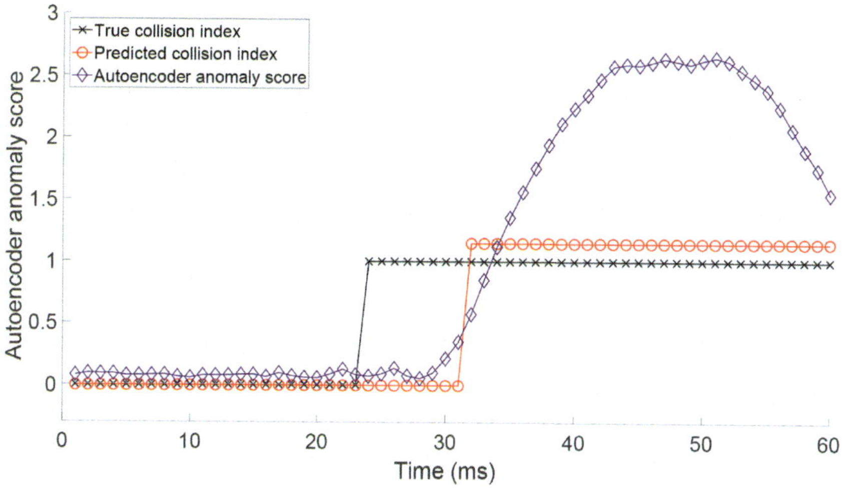

(a)

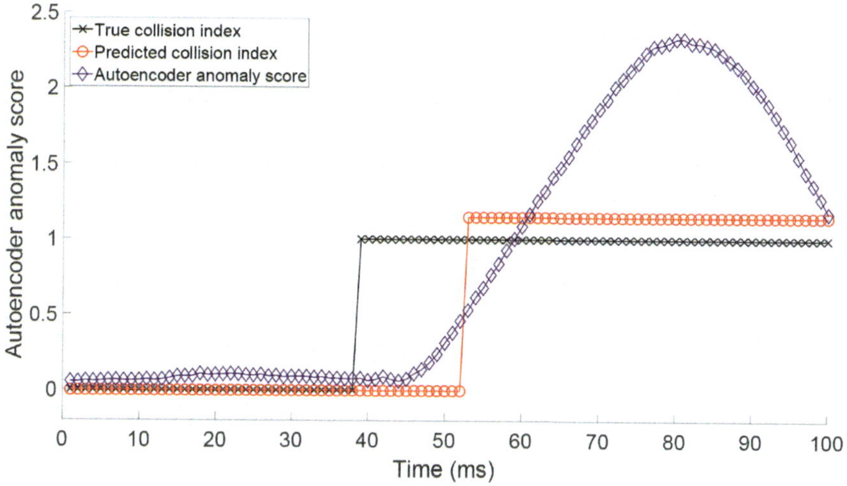

(b)

Fig. 5.9 MCD-AE detection results of **a** a hard collision and **b** a soft collision

a collision occurs. Figure 5.9a, b depict the anomaly score and its corresponding predicted collision index $cd_{AE}(g(x))$ of MCD-AE.

The number of detection failures and the average detection delay of the test collisions for the unsupervised learning-based detection methods are shown in Table 5.1. MCD-OCSVM successfully detects 99.65% of the hard collisions (2 failures out of 578 collisions) with an average detection delay of 15.94 ms and 97.61% of the soft collisions (5 failures out of 209 collisions) with an average detection delay of 41.83 ms. MCD-AE successfully detects 99.83% of the hard collisions (1 failure out

Table 5.1 The detection performance of the unsupervised learning-based detection methods

Methods	MCD-OCSVM		MCD-AE	
Collision type	Hard	Soft	Hard	Soft
Number of detection failures	2/578	5/209	1/578	5/209
Average detection delay (ms)	15.94	41.83	12.15	34.01
Number of false alarms	0/62.4 min		0/62.4 min	
Computation time (μs)	69.74		6.73	

of 578 collisions) with an average detection delay of 12.15 ms and 97.61% of the soft collisions (5 failures out of 209 collisions) with an average detection delay of 34.01 ms. Since the detection thresholds are set to produce a precision of 1, false alarms do not occur for 62.4 min.

The reason for failing to detect two hard collisions is that we do not use the observer outputs of joints 5 and 6 as inputs (5.9), (5.12) to reduce the computation burden; the two hard collisions are applied on link 5 but cause a noticeable change only in the observer output of joint 5 while causing very small changes in joints 1–4. On the other side, the detection failure of five soft collisions is due to small and slow changes in the observer outputs, where the five soft collisions require a time longer than t_{max} to reach their peak contact forces that are considerably small. However, the collisions the methods fail to detect are permissible to humans since the collisions have peak contact forces smaller than 65 N, which are then below the maximum permissible quasi-static/transient contact forces for whole human body regions described in Technical Specification ISO/TS 15066 [23].

The unsupervised learning-based detection methods are capable of accurately detecting a variety of hard and soft collisions (Fig. 4.9; Tables 4.3 and 4.4) without making any false alarms, validating that the methods can successfully compensate for the uncertain dynamic effects including the unmodeled friction. Among the collision detection methods for robot manipulators reviewed so far, some detection methods may present faster detection than the described unsupervised learning-based methods. However, the described methods clearly have advantages over those faster-detecting methods, e.g., not requiring additional external sensors or joint torque sensors, model uncertainty (including friction) compensation, easy detection sensitivity adjustment, and requiring only the collision-free data for training.

Table 5.2 Detection performance comparison between the unsupervised and supervised learning-based detection methods

Learning type	Training data (collision types)	Collision type	Hard collisions		Soft collisions	
		Method	Detection failures	Detection delay (ms)	Detection failures	Detection delay (ms)
Unsupervised	**Free**	MCD-OCSVM	2/578	15.94	5/209	41.83
		MCD-AE	1/578	12.15	5/209	34.01
Supervised	Free, Hard	MCD-CNN-Hard	3/578	12.47	**39/209**	**36.47**
	Free, Soft	MCD-CNN-Soft	**7/578**	**16.71**	2/209	27.75
	Free, Hard, Soft	MCD-CNN	**0/578**	**13.14**	**2/209**	**26.91**

5.4.3.3 Comparison with Supervised Learning-Based Methods

We now compare the detection performance between the unsupervised and supervised learning-based detection methods. The learning algorithm we consider for the supervised learning-based detection methods is 1-D CNN, which showed good detection performance in Chap. 4 (the 1-D CNN-based method MCD-CNN successfully detected a variety of hard and soft collisions). In addition, we also train two 1-D CNNs that have the same network architecture as MCD-CNN, one with only hard collision data (MCD-CNN-Hard), and the other with only soft collision data (MCD-CNN-Soft).

The required collision types for training, the number of detection failures, and the average detection delay for the unsupervised and supervised learning-based detection methods are shown in Table 5.2. MCD-CNN-Hard exhibits a similar level of detection performance to the unsupervised learning-based detection methods for hard collisions (99.48% detection rate with 12.47 ms average detection delay) but shows multiple detection failures for soft collisions (81.34% detection rate with 36.47 ms average detection delay). On the other hand, MCD-CNN-Soft shows better detection performance for soft collisions than the unsupervised learning-based detection methods (99.04% detection rate with 27.75 ms average detection delay) while showing slightly lower detection performance for hard collisions (98.79% detection rate with 16.71 ms average detection delay).

As mentioned in Sect. 5.1, supervised learning-based detection methods have the limitation in which only the scenarios that are learned as collisions can be robustly detected; while MCD-CNN-Hard exhibits multiple detection failures for soft collisions and MCD-CNN-Soft shows relatively low detection performance for hard collisions, MCD-OCSVM and MCD-AE detect both types of collision robustly. This implies that by using unsupervised anomaly detection algorithms, a variety of hard and soft collisions can be accurately detected without the knowledge of every possible type of collision that can occur, effectively overcoming the second limitation of supervised learning-based detection methods.

Compared to MCD-CNN described in Sect. 4.4.2, MCD-CNN shows better detection performance than the unsupervised learning-based detection methods, which is an expected result since MCD-CNN is trained with both hard and soft collisions. However, the robot may be damaged during the collision data collection and it is impractical to collect collision data for each of the mass-produced industrial robots. The unsupervised learning-based detection methods are therefore more practical for mass-produced industrial robots, requiring only the collision-free motion data for training.

5.4.3.4 Real-Time Inference

For real-time inference verification, we check the average computation times of the described methods. The unsupervised learning-based detection algorithms are implemented in C++ on a PC with an Intel Core i7-4770 CPU at 3.4 GHz and we measure the average computation times over 12,000 iterations. As shown in Table 5.1, an average computation time of 69.74 μs is measured for MCD-OCSVM, and 6.73 μs is measured for MCD-AE. For the robot to operate in real-time, the whole loop of the robot operating system should run within 1 ms, where the required computation times of the methods are significantly smaller than the maximum allowable inference time. The computation time of MCD-OCSVM depends on the dimension of the input feature (64-dimensional) and on the number of support vectors (737 support vectors). The computation time of MCD-AE depends on the number of layers (5 layers) and on the dimension of each layer (64-32-6-32-64).

5.4.4 Generalization Capability Analysis

In this section, similar to the supervised learning-based detection methods described in Chap. 4, we show the generalization capability of the unsupervised learning-based detection methods for the following two cases: (1) generalization to small changes in the robot dynamics model and random noise added to the motor current measurements and (2) generalization to an unseen payload of known mass. In Appendix, the generalization capability to large changes in the robot mass is also analyzed for the unsupervised learning-based detection methods.

5.4.4.1 Generalization to Small Perturbations

To roughly simulate the aging of the robot and the motor sensors, we decrease the mass matrix $M(q)$ by 10%, increase the Coriolis matrix $C(q, \dot{q})$ and the gravity vector $g(q)$ by 10%, and add white Gaussian noise with a signal-to-noise ratio of 20 dB to the motor torque measurement $\tau_m = K_i i_m$.

Table 5.3 The generalization capability of the unsupervised learning-based detection methods to small perturbations in the dynamics model and sensor measurements

Methods	MCD-OCSVM		MCD-AE	
Collision type	Hard	Soft	Hard	Soft
Change of number of detection failures	2/578 ↓ 3/578	5/209 ↓ 5/209	1/578 ↓ 1/578	5/209 ↓ 6/209
Change of average detection delay (ms)	15.94 ↓ 16.72	41.83 ↓ 45.23	12.15 ↓ 12.18	34.01 ↓ 35.29
Number of false alarms	0/62.4 min		0/62.4 min	

To increase the robustness to false alarms, we increase the absolute value of the detection threshold of MCD-OCSVM by 4%, in which the original threshold $\beta = -0.382$ is now decreased to $\beta = -0.397$ (the detection threshold of MCD-AE γ is not changed). Table 5.3 shows the detection performance of the unsupervised learning-based detection methods when small perturbations are applied to the dynamics model and sensor measurements, which is then compared with the detection performance when no perturbations are applied (Table 5.1). Compared to when no perturbations are applied, MCD-OCSVM fails to detect 1 more hard collision (3 detection failures), while showing the same detection failure rate for soft collisions (5 detection failures). The average detection delays are slightly increased for both hard (from 15.94 to 16.72 ms) and soft collisions (from 41.83 to 45.23 ms). MCD-AE shows the same detection failure rate as when no perturbations are applied for hard collisions (1 detection failure) while failing to detect 1 more soft collision (6 detection failures). The average detection delay is increased from 12.15 to 12.18 ms for hard collisions and from 34.01 to 35.29 ms for soft collisions. Both methods produce no false alarms. The results imply that the unsupervised learning-based detection methods can generalize to small perturbations applied to the dynamics model and random noise added to the motor current measurements, like the supervised learning-based detection methods.

5.4.4.2 Generalization to an Unseen Payload

We now show that the unsupervised learning-based detection methods can generalize to an unseen payload with its mass known. Similar to the supervised learning-based detection methods, we train a new OC-SVM model and a new autoencoder with only

Table 5.4 The generalization capability of MCD-OCSVM to an unseen payload (trained only with payload 0 and 3.3 kg data, and tested for all payload data)

Payload	0 kg, 3.3 kg		5.0 kg	
Detection threshold (β_1, β_2)	−0.3302		−0.3518	
Collision type	Hard	Soft	Hard	Soft
Number of detection failures	1/399	6/160	1/179	1/49
Average detection delay (ms)	16.14	39.39	17.02	56.54
Number of false alarms	0/41.7 min		0/20.7 min	

payload 0 and 3.3 kg data and test the detection performance for all three types of payload. The new OC-SVM model is trained with a total of 60,000 collision-free training samples (feature vectors), 30,000 for each payload. The input feature and kernel function hyperparameters are the same as those of the previous MCD-OCSVM. The new autoencoder is trained with a total of 80,000 collision-free training samples, 40,000 for each payload. The input feature and network architecture hyperparameters are the same as those of the previous MCD-AE.

Since the new MCD-OCSVM is trained only with payload 0 and 3.3 kg data, we use two detection threshold values: $\beta_1 \in \mathbb{R}$ for the trained cases (when there is no payload or 3.3 kg payload attached), and $\beta_2 \in \mathbb{R}$ for the unseen case (when 5.0 kg payload is attached). β_1 is set to the minimum value of $f(x)$ computed over the collision-free motions when no payload and 3.3 kg payload are attached, and β_2 is set in the same way as β_1, this time with the collision-free motions when 5.0 kg payload is attached. Table 5.4 shows the two detection thresholds $\beta_1 = -0.3302$ and $\beta_2 = -0.3518$, and the detection performance of the new MCD-OCSVM. For the trained cases, the new MCD-OCSVM successfully detects 99.75% of the hard collisions (1 failure out of 399 collisions) with an average detection delay of 16.14 ms and 96.25% of the soft collisions (6 failures out of 160 collisions) with an average detection delay of 39.39 ms. For the unseen case, 99.44% of the hard collisions (1 failure out of 179 collisions) are successfully detected with an average detection delay of 17.02 ms, and 97.96% of the soft collisions (1 failure out of 49 collisions) are successfully detected with an average detection delay of 56.54 ms. Since β_1 and β_2 are set to the minimum values of $f(x)$ computed over the collision-free motions, false alarms do not occur.

Similar to MCD-OCSVM, we use two detection thresholds for the new MCD-AE, $\gamma_1 \in \mathbb{R}$ for the trained cases and $\gamma_2 \in \mathbb{R}$ for the unseen case. The detection thresholds are set to the maximum values of $g(x)$ computed over the collision-free motions, when no payload and 3.3 kg payload are attached for γ_1 and when 5.0 kg payload is attached for γ_2. Table 5.5 shows the two detection thresholds $\gamma_1 = 0.4572$ and $\gamma_2 = 0.4959$, and the detection performance of the new MCD-AE. For the trained cases, the

Table 5.5 The generalization capability of MCD-AE to an unseen payload (trained only with payload 0 and 3.3 kg data, and tested for all payload data)

Payload	0 kg, 3.3 kg		5.0 kg	
Detection threshold (γ_1, γ_2)	0.4572		0.4959	
Collision type	Hard	Soft	Hard	Soft
Number of detection failures	0/399	3/160	1/179	2/49
Average detection delay (ms)	10.93	33.55	12.19	45.62
Number of false alarms	0/41.7 min		0/20.7 min	

new MCD-AE fails to detect 3 soft collisions (1.88% failure rate, average detection delay 33.55 ms) and successfully detects all hard collisions (average detection delay 10.93 ms). For the unseen case, the new MCD-AE fails to detect 1 hard collision (0.56% failure rate, average detection delay 12.19 ms) and 2 soft collisions (4.08% failure rate, average detection delay 45.62 ms).

For both the new MCD-OCSVM and the new MCD-AE, to eliminate all the false positives that occur during the unseen case, detection thresholds with higher absolute values are required compared to the trained cases ($\beta_1 = -0.3302 \rightarrow \beta_2 = -0.3518$, $\gamma_1 = 0.4572 \rightarrow \gamma_2 = 0.4959$). Nevertheless, the unsupervised learning-based detection methods that are trained only with payload 0 and 3.3 kg data can still accurately detect collisions in the case of payload 5.0 kg data as shown in Tables 5.4 and 5.5, successfully generalizing to an unseen payload.

5.4.5 *Discussion*

We have verified the detection performance of the unsupervised learning-based detection methods with the test dataset we have collected. The experimental results show that the unsupervised learning-based methods are able to robustly detect a variety of hard and soft collisions with very light computation, validating that uncertain dynamic effects (e.g., unmodeled or uncertain parameters in the robot dynamics model, measurement noise) including the unmodeled friction can be successfully compensated by using unsupervised anomaly detection algorithms. Moreover, the unsupervised learning-based methods require only one constant detection threshold to be tuned. The generalization capability of the unsupervised learning-based methods has also been verified for the collected dataset.

Although the supervised learning-based methods show better detection performance than the unsupervised learning-based methods (which is an expected result), the unsupervised learning-based methods require only the data for collision-free motions for training, which makes them more practical for mass-produced industrial robots. We also show that a variety of hard and soft collisions can be accurately detected without the knowledge of every possible type of collision that can occur by using unsupervised anomaly detection algorithms.

Comparing the two unsupervised learning-based methods, the autoencoder-based method shows marginally better performance than the OC-SVM-based method with significantly less computation required, being the preferred choice in general. However, in our experiments, the performance of OC-SVM was more stable than that of an autoencoder with respect to the selection of hyperparameters. Therefore, we would recommend the OC-SVM-based method for practitioners without sufficient resources for hyperparameter tuning.

References

1. Sharkawy, A.-N., Koustoumpardis, P.N., Aspragathos, N.: Neural network design for manipulator collision detection based only on the joint position sensors. Robotica, **38**(10), 1737–1755 (2020)
2. Sharkawy, A.-N., Koustoumpardis, P.N., Aspragathos, N.: Human–robot collisions detection for safe human–robot interaction using one multi-input–output neural network. Soft Comput. **24**(9), 6687–6719 (2020)
3. Heo, Y.J., Kim, D., Lee, W., Kim, H., Park, J., Chung, W.K.: Collision detection for industrial collaborative robots: a deep learning approach. IEEE Robot. Autom. Lette. **4**(2), 740–746 (2019)
4. Zhang, Z., Qian, K., Schuller, B.W., Wollherr, D.: An online robot collision detection and identification scheme by supervised learning and Bayesian decision theory. IEEE Trans. Autom. Sci. Eng. (2020)
5. Kim, D., Lim, D., Park, J.: Transferable collision detection learning for collaborative manipulator using versatile modularized neural network. IEEE Trans. Robot. (2021)
6. Park, K.M., Kim, J., Park, J., Park, F.C.: Learning-based real-time detection of robot collisions without joint torque sensors. IEEE Robot. Autom. Lett. **6**(1), 103–110 (2020)
7. Park, K.M., Park, Y., Yoon, S., Park, F.C.: Collision detection for robot manipulators using unsupervised anomaly detection algorithms. IEEE/ASME Trans. Mechatron. (2021)
8. Tian, X., Fan, J., Fang, Q., Zhu, Y., Zhao, J.: A new robot collision detection method: a modified nonlinear disturbance observer based-on neural networks. J. Intell. Fuzzy Syst. **38**(1), 175–186 (2020)
9. Lim, D., Kim, D., Park, J.: Momentum observer-based collision detection using LSTM for model uncertainty learning. In: 2021 IEEE International Conference on Robotics and Automation (ICRA), pp. 4516–4522. IEEE (2021)
10. Pan, S.J., Yang, Q.: A survey on transfer learning. IEEE Trans. Knowl. Data Eng. **22**(10), 1345–1359 (2009)
11. Wu, H., Zhao, J.: Fault detection and diagnosis based on transfer learning for multimode chemical processes. Comput. Chem. Eng. **135**, 106731 (2020)
12. Wang, Y., Wu, D., Yuan, X.: LDA-based deep transfer learning for fault diagnosis in industrial chemical processes. Comput. Chem. Eng. **140**, 106964 (2020)
13. Shao, S., McAleer, S., Yan, R., Baldi, P.: Highly accurate machine fault diagnosis using deep transfer learning. IEEE Trans. Ind. Inform. **15**(4), 2446–2455 (2018)
14. Xiao, D., Huang, Y., Zhao, L., Qin, C., Shi, H., Liu, C.: Domain adaptive motor fault diagnosis using deep transfer learning. IEEE Access **7**, 80937–80949 (2019)

15. Kim, S., Kim, W., Noh, Y.-K., Park, F.C.: Transfer learning for automated optical inspection. In: 2017 International Joint Conference on Neural Networks (IJCNN), pp. 2517–2524. IEEE (2017)
16. Kim, S., Noh, Y.-K., Park, F.C.: Efficient neural network compression via transfer learning for machine vision inspection. Neurocomputing **413**, 294–304 (2020)
17. Narukawa, K., Yoshiike, T., Tanaka, K., Kuroda, M.: Real-time collision detection based on one class SVM for safe movement of humanoid robot. In: 2017 IEEE-RAS 17th International Conference on Humanoid Robotics (Humanoids), pp. 791–796. IEEE (2017)
18. Chen, T., Liu, X., Xia, B., Wang, W., Lai, Y.: Unsupervised anomaly detection of industrial robots using sliding-window convolutional variational autoencoder. IEEE Access **8**, 47072–47081 (2020)
19. Sindhwani, V., Sidahmed, H., Choromanski, K., Jones, B.: Unsupervised anomaly detection for self-flying delivery drones. In: 2020 IEEE International Conference on Robotics and Automation (ICRA), pp. 186–192. IEEE (2020)
20. Park, J.-H., Shanbhag, S., Chang, D.E.: Model-free unsupervised anomaly detection of a general robotic system using a stacked LSTM and its application to a fixed-wing unmanned aerial vehicle. In: 2022 IEEE/RSJ International Conference on Intelligent Robots and Systems (IROS), pp. 4287–4293. IEEE (2022)
21. Quinlan, M., Chalup, S., Middleton, R.: Application of SVMs for colour classification and collision detection with AIBO robots. Adv. Neural Inform. Process. Syst. **16** (2003)
22. Çatal, O., Leroux, S., De Boom, C., Verbelen, T., Dhoedt, B.: Anomaly detection for autonomous guided vehicles using Bayesian surprise. In: 2020 IEEE/RSJ International Conference on Intelligent Robots and Systems (IROS), pp. 8148–8153. IEEE (2020)
23. ISO/TS 15066:2016: Robots and Robotic Devices—Collaborative Robots. International Organization for Standardization (2016)

Chapter 6
Conclusion

6.1 Summary and Discussion

This book has presented a comprehensive survey of collision detection methods for robot manipulators, with a focus on four learning-based detection methods, two using supervised learning algorithms and the other two based on unsupervised anomaly detection. Existing collision detection methods can be loosely categorized into model-free, model-based, and learning-based approaches, with overlaps among the categories. The fundamentals of collision detection handling have also been covered in a systematic fashion: beginning with a characterization of collision types and detection sensitivity adjustment, we covered collision detection performance criteria, from conventional measures such as detection delay to learning-related criteria such as generalization capability.

- **Fundamentals of Collision Detection**
  In Chap. 2, we first briefly reviewed dynamic models for both rigid joint and flexible joint robot manipulators that are widely used in model-based collision detection methods. Collision types, detection sensitivity adjustment, and collision detection performance criteria have also been summarized. Collisions are categorized into two types: (i) hard collisions, which are transient impact collisions having a duration time on the order of 10–100 ms, and (ii) soft collisions, which are quasi-static collisions that have a duration time in the range of 100–700 ms. Most collision detection methods allow for a means of adjusting detection sensitivities, either in the form of a vector or scalar adjustment parameter. We discussed the relation between the detection threshold and sensitivity, and also introduced a simple method to find the optimal threshold. Collision detection performance criteria include PRAUC, detection delay, the number of detection failures and false alarms, computation time, and generalization capability. The global performance of a detection method can be indicated by the PRAUC, while detection performance with a fixed detection threshold is evaluated with the detection delay and

K. M. Park and F. C. Park, *Collision Detection for Robot Manipulators: Methods and Algorithms*, Springer Tracts in Advanced Robotics 155,
https://doi.org/10.1007/978-3-031-30195-7_6

the number of detection failures and false alarms. The computation time must be taken into account for stable real-time computation. For learning-based detection methods, generalization capability is a particularly relevant performance criterion.

- **Model-Free and Model-Based Methods**
  In Chap. 3, we covered various model-free and model-based detection methods. Model-free methods typically rely on additional external sensors—tactile surface sensors, three-axis accelerometers, microphones, or cameras. These methods show desirable features not only for detection but also for isolation, identification, and classification. However, from both a cost and performance perspective, model-based methods that rely on a robot dynamics model together with joint signals have overall been more successful than model-free methods. For the model-based methods, we first described four basic methods based on the following key concepts: energy observer, direct estimation, velocity observer, and momentum observer. Among the four methods, it is generally agreed for now that the momentum observer-based method is the most accurate, reliable, and cost-effective.
  We have also introduced several approaches to deal with frictional torque in the joints, including both analytical and learning-based models for friction as well as using additional sensors. Collision detection under the special case of the robot performing cyclic motions has also been discussed. Finally, after introducing some alternative model-based methods (improving the performance of the momentum observer, dynamics model compensating methods, and joint acceleration estimation), extensions to multiple arms and to other end-effector types have been discussed.

- **Supervised Learning-Based Detection Methods**
  In Chap. 4, we have provided a survey of recent supervised learning-based detection methods, summarizing the key features, advantages, and disadvantages of the introduced methods. Supervised learning-based methods require true labels for the training data, which include binary true collision index and torque values. These methods can be categorized into two types based on the required collision data: those that require data for both collision and collision-free motions of the robot, and those that require only data for collision-free motions. If only the data for collision-free motions is available, friction learning methods introduced in Chap. 3 can be used.
  When collision data is available for training, one of the following three approaches can be used depending on the output of the utilized learning algorithm: (i) estimating the external joint torque without full dynamics; (ii) estimating the binary predicted collision index with an end-to-end network; (iii) computing a scalar collision prediction score. Two supervised learning-based detection methods were discussed in some detail, one based on SVMR and the other using a 1-D CNN. Based on experimental studies for detection performance and generalization capability, it was shown that these methods can accurately detect a wide range of hard and soft collisions in real-time using a light network, compensating for uncertainties in the model parameters.

- **Methods for Enhancing Collision Learning Practicality**
  In Chap. 5, we introduced several approaches to overcome the limitations of supervised learning-based detection methods and to enhance the practicality of learning, i.e., making them easier to be replicated for mass-produced industrial robots. These approaches include methods that use unsupervised anomaly detection—learning collision-free motions to detect anomalous monitoring signals during collisions, and a transfer learning-based scheme in which a pre-trained detection network is transferred from the source robot to the target robot without the need for collision data from the target robot.
  We further classified the introduced learning-based detection methods according to several criteria: the number of detection thresholds, applicability to general non-periodic motions, and whether or not the method requires a full dynamics model, collision training data, and joint torque sensors. We also described in some detail the implementation methodology and verified the detection performance of two unsupervised learning-based detection methods, one using an OC-SVM and the other based on an autoencoder. It was verified that the unsupervised anomaly detection methods can detect a variety of hard and soft collisions accurately, without prior knowledge of every possible type of collision that can occur.

It should be noted that the learning-based detection methods introduced in this book (and likely other learning-based methods that use the observer output as input) can, with minor modifications, be applied to other robot manipulator designs in addition to the manipulators addressed in the experiments, as long as the joint signals and a robot dynamics model are available. For a manipulator with a different number of joints, the only required variation is the selection of the observer outputs to be used as the input for the learning algorithm, e.g., using the observer outputs of joints 1–6 as the input in the case of a 7-DOF manipulator. By replacing the momentum observer for rigid joint robots with the link-side or the total momentum observer, the methods can also be applied to manipulators with flexible joints.

6.2 Future Work

For learning-based detection methods, both the learning algorithm (network architecture) and the quality and quantity of the training data are important factors to be considered for achieving high detection performance. This naturally raises the question of what is more important, the quantity and quality of the data or the algorithm? To this question there is no general answer; rather, it is important to analyze, on a case-by-case basis, the effect each has on overall detection performance.

Also, for the safe operation of collaborative robot manipulators, not only the detection phase but also the isolation, identification, and classification phases are important. Applying learning-based approaches to the next phases remains an important topic for future work.

Appendix A
Collision Classification

A.1 SVM-Based Classification of Detected Collisions

We expand the SVMR-based detection method (MCD-SVMR) to the classification of hard and soft collisions using a support vector machine (SVM), which is a well-known binary classifier. The SVM is trained to classify the input signal features (4.21) that are signaled as collisions into hard and soft collisions. Since MCD-SVMR produces no false alarms (input features signaled as collisions are all true collision situations), binary classification (no collision-free class) is sufficient. A block diagram that illustrates the whole process of detection and classification is shown in Fig. A.1.

For the SVM model, we use the RBF kernel (5.10) with its hyperparameter set to $\sigma = 1/\sqrt{2}$, and set the hyperparameter of the SVM as $C = 100$ (we refer to [1] for details of the hyperparameter). The SVM model is trained with 418 hard collision-labeled samples (labeled as positives) and 627 soft collision-labeled samples (labeled as negatives), where sequential minimal optimization is used for optimization. The

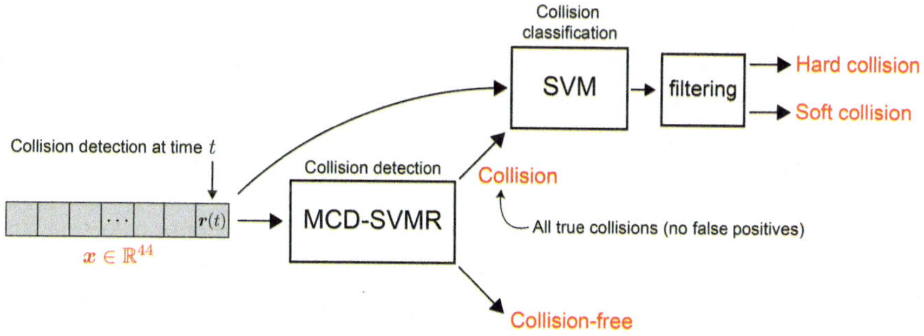

Fig. A.1 The SVMR-based detection and the SVM-based classification

K. M. Park and F. C. Park, *Collision Detection for Robot Manipulators: Methods and Algorithms*, Springer Tracts in Advanced Robotics 155,
https://doi.org/10.1007/978-3-031-30195-7

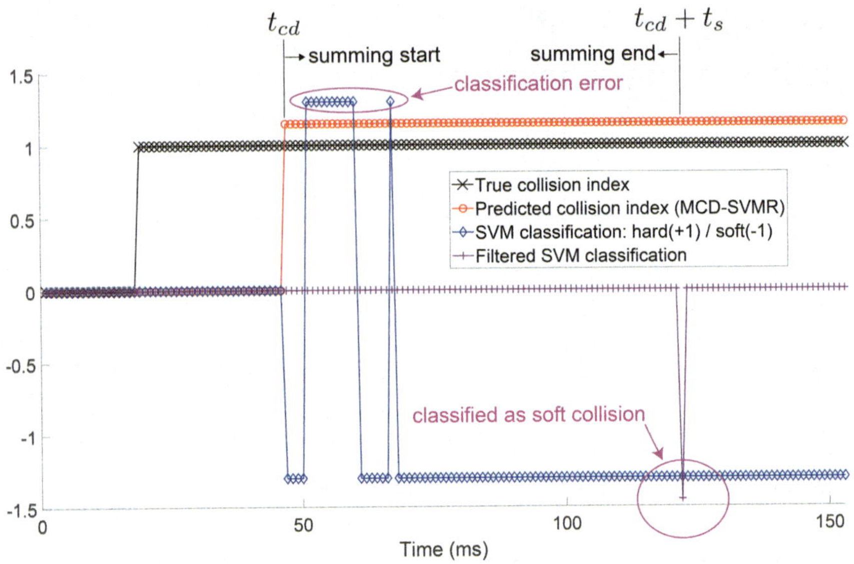

Fig. A.2 The SVM output and its filtered classification result when a soft collision occurs

output value $s(x)$ of the trained SVM model is a binary scalar value that is $+1$ for hard collision-labeled inputs and -1 for soft collision-labeled inputs.

However, when a soft collision occurs, the SVM model may misclassify the collision at several time steps especially at the beginning of the collision as shown in Fig. A.2, since the soft collision may momentarily show an estimated external joint torque profile similar to that of a hard collision. Therefore, we apply a simple output filtering to the SVM output $s(t)$. When a collision is detected by MCD-SVMR at time t_{cd}, we sum the SVM output for t_s ms, calculated as $s_{filtered}(t_{cd} + t_s) = \sum[s(t_{cd}), s(t_{cd} + t_s)]$. If $s_{filtered}(t_{cd} + t_s)$ is positive, then the detected collision is classified as a hard collision, and as a soft collision if negative. In our case, we set $t_s = 75$ ms. Figure A.2 shows the SVM output and the filtered classification result of a soft collision.

We test the SVM-based classifier for the test collisions that are detected by MCD-SVMR—577 hard collisions and 208 soft collisions (1 detection failure for each type). The classifier correctly classifies 547 out of 577 hard collisions (94.8 % success rate) and 189 out of 208 soft collisions (90.9 % success rate). The confusion matrix for the classification result is shown in Fig. A.3.

Although extra time is required for classifying compared to only detecting (the total time required for classification is the sum of the detection delay introduced by MCD-SVMR and t_s introduced by the classifier filter), we validate that the detection methods we have developed can easily be expanded to the classification of hard and soft collisions.

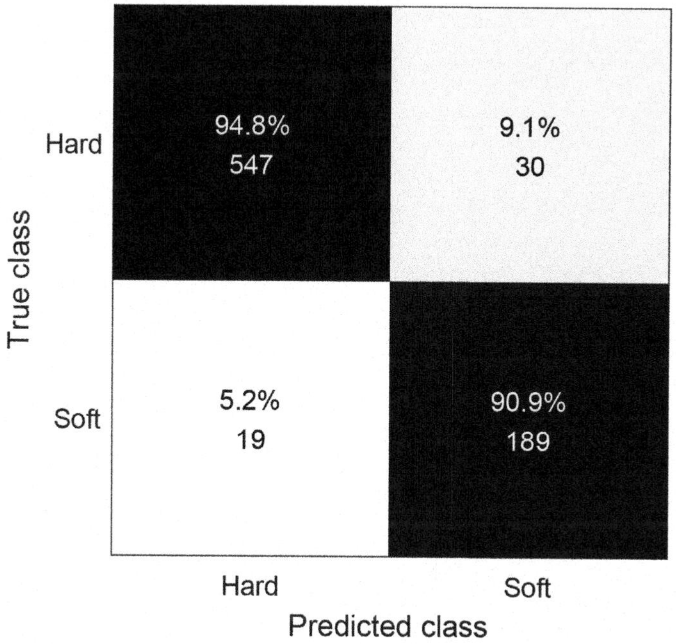

Fig. A.3 The confusion matrix of the SVM-based classification result

Reference

1. Burges, C.J.C.: A tutorial on support vector machines for pattern recognition. Data Min. Knowl. Discovery **2**(2), 121–167 (1998)

Appendix B
Alternative Supervised Learning-Based Detection Methods

B.1 Direct Estimation-Based Detection Methods

We describe here two alternative supervised learning-based detection methods that take as inputs the external joint torque estimated by direct estimation. The main monitoring signals are the active motor torque $\tau_m(t)$ and the *reference torque* $\tau_{ref}(t) \in \mathbb{R}^n$ given by

$$\tau_{ref} = \hat{M}(q)\ddot{q}_d + \hat{C}(q, \dot{q})\dot{q} + \hat{g}(q), \tag{B.1}$$

where $\ddot{q}_d(t) \in \mathbb{R}^n$ is the desired joint acceleration and $\hat{M}$, $\hat{C}$, and $\hat{g}$ denote the nominal model parameters. Since joint acceleration $\ddot{q}(t)$ is not directly measured and numerical differentiation of the measured joint velocity is inherently noisy, we instead use the desired joint acceleration $\ddot{q}_d(t)$. Under ideal conditions where $\hat{M} = M$, $\hat{C} = C$, and $\hat{g} = g$, the estimated external joint torque $\hat{\tau}_{ext} \in \mathbb{R}^n$ is calculated as

$$\begin{aligned} \hat{\tau}_{ext} &= \tau_{ref} - \tau_m \\ &= \hat{M}\ddot{q}_d + \hat{C}\dot{q} + \hat{g} - (M\ddot{q} + C\dot{q} + g + \tau_F - \tau_{ext}) \\ &= M(\ddot{q}_d - \ddot{q}) + \tau_{ext} - \tau_F \\ &\approx \tau_{ext} - \tau_F, \end{aligned} \tag{B.2}$$

where the robot dynamics with external joint torque (2.2) is used. We use the absolute value of this directly estimated external joint torque $\hat{\tau}_{ext}$ as inputs for the SVMR and the 1-D CNN.

The input feature vector for the SVMR is designed by replacing the momentum observer output r in (4.21) with the estimated external torque $\hat{\tau}_{ext}$. The time sampling parameters and the feature labeling parameters are the same as MCD-SVMR. An SVMR model using the SRQ kernel function (4.22) with the same hyperparameters as MCD-SVMR is trained with 1,890,105 features including 105 collision-labeled features. The number of support vectors of the trained SVMR model is 385.

K. M. Park and F. C. Park, *Collision Detection for Robot Manipulators: Methods and Algorithms*, Springer Tracts in Advanced Robotics 155,
https://doi.org/10.1007/978-3-031-30195-7

Table B.1 The detection performance of the supervised learning-based detection methods that use direct estimation

Methods	SVMR-based		1-D CNN-based	
Collision type	Hard	Soft	Hard	Soft
Detection failures	4/578	1/209	5/578	4/209
Average detection delay	20.43 ms	34.92 ms	14.96 ms	32.03 ms
False alarms	0/62.4 min		0/62.4 min	

The input feature matrix for the 1-D CNN is designed by replacing the momentum observer output r in (4.24) with the estimated external torque $\hat{\tau}_{ext}$. The time sampling parameters and the 1-D CNN network architecture are the same as those of MCD-CNN (Fig. 4.16, Table 4.5), and the network is trained with 8,036,355 features composed of 16,040 collision-labeled features and 8,020,315 collision-free-labeled features.

Table B.1 shows the detection performance (detection delay and the number of detection failures) of the direct estimation-based detection methods. The detection threshold for the SVMR-based method is set to 0.021. For both SVMR-based and 1-D CNN-based methods, a continuous filter with $t_c = 7$ ms is applied. The directly estimated external joint torque $\hat{\tau}_{ext}(t)$ (B.2) exhibits more noise than the momentum observer output $r(t)$ (4.14), requiring heavier output filtering than using $r(t)$ in order to eliminate all the false alarms. This leads to increments in detection failure rates and average detection delays.

B.2 Model-Free Detection Methods

We describe here two model-free supervised learning-based detection methods that require only the motor current measurements as input. For the model-free methods, an input filtering method to reduce the measurement noise is introduced.

To better compensate for the noise in the motor current measurements, we apply the simple moving average filtering to the motor current measurement $i_m(t)$. The filtered measurement at time t is computed by the mean value of the past t_{MA} ms measurements, $[i_m(t - t_{MA}), i_m(t)]$. The moving average filter with $t_{MA} = 20$ ms is applied to both the SVMR-based and 1-D CNN-based methods.

The input feature vector for the SVMR is designed by replacing the momentum observer output r in (4.21) with the motor current value i_m. The time sampling parameters and the feature labeling parameters are the same as MCD-SVMR. An SVMR model using the SRQ kernel function (4.22) with the same hyperparameters as MCD-SVMR is trained with 1,920,120 features including 120 collision-labeled features. The number of support vectors of the trained SVMR model is 351.

Table B.2 The detection performance of the model-free supervised learning-based detection methods

Methods	SVMR-based		1-D CNN-based	
Collision type	Hard	Soft	Hard	Soft
Detection failures	4/578	11/209	9/578	13/209
Average detection delay	25.20 ms	37.81 ms	14.64 ms	32.27 ms
False alarms	0/62.4 min		0/62.4 min	

The input feature matrix for the 1-D CNN is designed by replacing the momentum observer output r in (4.24) with the motor current value i_m. The time sampling parameters and the 1-D CNN network architecture are the same as those of MCD-CNN (Fig. 4.16, Table 4.5), and the network is trained with 8,032,506 features composed of 12,191 collision-labeled features and 8,020,315 collision-free-labeled features. For both SVMR-based and 1-D CNN-based methods, a continuous filter with $t_c = 4$ ms is applied.

Table B.2 shows the detection performance of the model-free supervised learning-based detection methods. The detection threshold for the SVMR-based method is set to 0.024. Although detection of both hard and soft collisions without the dynamics model is realized by learning only the motor current measurements, the model-free methods show a relatively high detection failure rate, which is critical for safe human-robot collaborative operation.

Appendix C
Additional Generalization Capability Analysis

C.1 Generalization to Large Changes in the Dynamics Model

In this section, we roughly simulate the situations in which the supervised and unsupervised learning-based detection methods trained on *M0609* are used in robots with 50% smaller or 50% larger mass (including the payload) without any additional training: the dynamics model of the robot $M(q)$, $C(q, \dot{q})$, and $g(q)$ and the active motor torque τ_m are (1) decreased and (2) increased by 50%.

Table C.1 shows the adjusted detection threshold (α) and the detection performance of MCD-SVMR when large perturbations are applied to the dynamics model (50% decreased and 50% increased). MCD-SVMR fails to detect 8 hard collisions

Table C.1 The generalization capability of MCD-SVMR to large changes in the dynamics model

Dynamics model change	50% decrease		50% increase	
Detection threshold (α)	0.013		0.1491	
Collision type	Hard	Soft	Hard	Soft
Number of detection failures	8/578	10/209	76/578	10/209
Average detection delay	26.11 ms	46.92 ms	28.98 ms	59.89 ms
Number of false alarms	0/62.4 min		0/62.4 min	

Table C.2 The generalization capability of MCD-CNN to a large change in the dynamics model

Dynamics model change	50% decrease	
Output filter (t_c)	1 ms	
Collision type	Hard	Soft
Number of detection failures	26/578	42/209
Average detection delay	10.68 ms	57.32 ms
Number of false alarms	0/62.4 min	

K. M. Park and F. C. Park, *Collision Detection for Robot Manipulators: Methods and Algorithms*, Springer Tracts in Advanced Robotics 155,
https://doi.org/10.1007/978-3-031-30195-7

Table C.3 The generalization capability of MCD-OCSVM to a large change in the dynamics model

Dynamics model change	50% decrease	
Detection threshold (β)	−0.0949	
Collision type	Hard	Soft
Number of detection failures	15/578	25/209
Average detection delay	21.28 ms	55.60 ms
Number of false alarms	0/62.4 min	

Table C.4 The generalization capability of MCD-AE to large changes in the dynamics model

Dynamics model change	50% decrease		50% increase	
Detection threshold (γ)	0.2711		0.7556	
Collision type	Hard	Soft	Hard	Soft
Number of detection failures	5/578	11/209	1/578	5/209
Average detection delay	12.45 ms	39.54 ms	11.82 ms	29.66 ms
Number of false alarms	0/62.4 min		0/62.4 min	

(1.38% failure rate) and 10 soft collisions (4.78% failure rate) when the dynamics model is decreased by 50% and 76 hard collisions (13.15% failure rate) and 10 soft collisions (4.78% failure rate) when the dynamics model is increased by 50%, both showing much higher detection failure rates than when no perturbations are applied (lower than 1%). For MCD-CNN, when the dynamics model is increased, the network does not work properly, showing multiple detection failures and false alarms at the same time. When the dynamics model is decreased, as shown in Table C.2, the continuous output filtering is no longer needed ($t_c = 1$ ms) since no false alarms occur but fails to detect 26 hard collisions (4.5% failure rate) and 42 soft collisions (20.09% failure rate).

Table C.3 shows the adjusted detection threshold (β) and the detection performance of MCD-OCSVM when the dynamics model is decreased: MCD-OCSVM fails to detect 15 hard collisions (2.59% failure rate) and 25 soft collisions (11.96% failure rate). When the dynamics model is increased, MCD-OCSVM fails to correctly classify the inputs into collision and collision-free. Finally, Table C.4 shows the adjusted detection threshold (γ) and the detection performance of MCD-AE when large perturbations are applied to the dynamics model. When the dynamics model is decreased, MCD-AE fails to detect 5 hard collisions (0.87% failure rate) and 11 soft collisions (5.26% failure rate). On the other hand, when the dynamics model is increased, MCD-AE shows the same detection failure rate as when no perturbations are applied (1 hard collision and 5 soft collisions) and even shows faster detection speed (hard collisions: 12.15 ms → 11.82 ms, soft collisions: 34.01 ms → 29.66 ms).

From the simulation results, we can conclude that the learning-based detection methods should not be used in robots with large mass differences without additional training, except in the case of using MCD-AE in robots with larger mass.

The manufacturer's authorised representative in the EU is Springer Nature Customer Service Centre GmbH, Europaplatz 3, 69115 Heidelberg, Germany. If you have any concerns regarding our products, please contact ProductSafety@springernature.com

Printed and bound by CPI Group (UK) Ltd, Croydon, CR0 4YY
01/10/2026
02225877-0004